Instrumentation
Temperature

Instrumentation

Temperature

F. E. Doyle, M.INST.M.C.
Lecturer, Bolton Technical College, Lancashire

G. T. Byrom, M.INST.M.C.
Lecturer, Bolton Institute of Technology, Lancashire

Blackie
Glasgow . London

Blackie & Son Limited
5 Fitzhardinge Street,
London W.1.

Bishopbriggs Glasgow

Blackie & Son (India) Limited
103-5 Fort Street,
Bombay

© F. E. Doyle, G. T. Byrom 1970
First published 1970

All Rights Reserved. No part of this
publication may be reproduced, stored
in a retrieval system, or transmitted,
in any form or by any means, electronic,
mechanical, photocopying, recording or
otherwise, without the prior permission
of Blackie & Son Limited.

Printed in Great Britain by
Robert Cunningham & Sons Ltd
Longbank Works,
Alva,
Scotland

ISBN 0 216 87428 9
Dewey 620 78(1)

Contents

The Measurement of Temperature
International Temperature Scale ... 1

Part 1 Change of State
Temperature-sensitive Paints and Crayons ... 2

Part 2 Expansion-type Thermometers
2.1. *Bimetal Thermometers* ... 3
2.2. *Bimetal Thermostat* ... 5
    Example ... 6
2.3. *Liquid-in-glass Thermometers* ... 6
    Example ... 7
2.3.1. Liquids other than Mercury ... 8
2.3.2. Maximum and Minimum Thermometers ... 8
2.3.3. Calibration ... 8
2.4. *Liquid-in-metal Thermometers* ... 10
2.4.1. Bulbs ... 10
2.4.2. Capillaries ... 11
2.4.3. Ambient-temperature Errors ... 11
    Example ... 11
2.4.4. Ambient-temperature Compensation ... 12
2.4.5. Case Compensation ... 12
2.4.6. Other Errors ... 13
2.4.7. Other Filling Liquids ... 13
2.4.8. Refilling ... 13
2.4.9. Range ... 14
2.4.10. Pockets and Sheaths ... 14
2.5. *Temperature Regulators* ... 14

Part 3 Pressure-type Thermometers
3.1. *Gas-filled Thermometers* ... 16
    Example ... 16
3.1.1. Refilling ... 16
3.2. *Vapour Pressure Thermometers* ... 18
3.2.1. Errors ... 18
3.2.2. Overcoming Cross Ambient Effects ... 18
3.2.3. Refilling ... 18
    TABLE 1. PROPERTIES OF SOME FILLING LIQUIDS ... 20
3.2.4. Remote Transmission ... 20
3.2.5. General ... 20
3.2.6. Pipe Runs ... 20
    TABLE 2. SUMMARY OF MECHANICAL THERMOMETERS ... 22
City and Guilds Examination Questions ... 22

Part 4 Electrical Thermometers
4.1. *Resistance Thermometers* ... 23
4.1.1. Callendar—Van Dusen Equation ... 23
    Example ... 23
    Example ... 23
4.1.2. Resistance Thermometer Elements ... 23
4.1.3. Wire-wound Elements ... 23
4.1.4. Time Constant ... 24
4.1.5. Self Heating ... 24
4.1.6. Construction ... 24
4.1.7. Photo-etched Elements ... 24
4.1.8. Parasitic Phenomena ... 25
4.1.9. Thermistor Elements ... 25
4.1.10. N.T.C. Thermistor Characteristic ... 25
    Example ... 25
4.1.11. Linearization ... 26
4.1.12. Advantages and Disadvantages ... 26
4.1.13. Measuring Circuits ... 26
4.1.14. Principle of Operation ... 26
4.1.15. Lead Resistance ... 27
4.1.16. Three- and Four-Wire Circuits ... 27
    Example ... 28
4.1.17. Multipoint Installations ... 28
4.1.18. Differential Temperature ... 29
4.1.19. Use of Wheatstone Bridge ... 29
4.2. *Thermocouple Thermometers* ... 29
4.2.1. General Form of e.m.f./temperature curve ... 30
    Example ... 30
    Example ... 30
4.2.2. Peltier Effect ... 30
4.2.3. Thomson Effect ... 30
    Example ... 30
4.2.4. Law of Intermediate Temperatures ... 31
4.2.5. Law of Intermediate Metals ... 31
4.2.6. Thermoelectric Power ... 31
4.2.7. Thermocouple Leads ... 32
4.2.8. Cold-junction Reference ... 32
4.2.9. Resistance of Thermocouple Circuit ... 32
    Example ... 32
4.2.10. Cold-junction Compensation ... 34
4.2.11. The Millivolt Indicator ... 34
4.2.12. Errors and Corrections ... 34
4.2.13. The Potentiometric Circuit ... 34
4.2.14. Differential Temperature Measurement and Multipoint Installations ... 36
4.2.15. Choice of Thermocouple ... 36
4.2.16. Thermocouple Materials ... 36

	TABLE 3. INDUSTRIAL THERMOCOUPLE METALS	36
4.2.17.	Calibration	38
4.2.18.	Installation	38
4.2.19.	Automatic Self-balancing Systems	38
4.2.20.	Commercial Types of Measuring Instruments. The Tenzor	38
4.2.21.	Foster Potentiometric Recorder	38
4.2.22.	The Fielden 'Bikini'	38
	TABLE 4. SUMMARY OF ELECTRICAL THERMOMETERS—RESISTANCE AND THERMOCOUPLE	40
	Examination Questions	40

Part 5 Radiation Pyrometers

5.0.1.	Prévost's Theory	42
5.0.2.	Black-body Conditions	42
5.0.3.	Stefan-Boltzmann Law	42
5.1.	*Total-radiation Pyrometers*	42
5.1.1.	Lens Material	42
5.1.2.	Fixed-focus Tubes	42
5.1.3.	Varying-focus Tubes	44
5.1.4.	Extension of Range	44
5.2.	*Optical Pyrometers*	44
5.2.1.	Disappearing-filament Pyrometer (Variable-intensity type)	44
5.2.2.	Wedge-type Pyrometer (Constant Intensity)	46
5.3.	*Photoelectric Pyrometers*	46
5.3.1.	Fibre-optic Methods	46
	Examination Questions	46
	TABLE 5. SUMMARY OF RADIATION PYROMETERS	47

Part 6 Calibration and Testing Methods

6.1.	*Fixed-point methods*	48
6.1.1.	Ice and Steam Points	48
6.1.2.	Sulphur Point	48
6.2.	*Freezing-point Methods*	48
6.3.	*Melting-point of Wire Method*	48
6.4.	*Comparison Methods*	50
6.5.	*Calibration above $1063°C$*	50
6.5.1.	Secondary Standards of Calibration	50
6.5.2.	Calibration of Radiation Pyrometers by the Melting-point of Refractory Cones	50
	City and Guilds of London Examination Questions	50
	INDEX	53

Acknowledgements

The information supplied by various manufacturers is acknowledged with thanks; they are mentioned in the appropriate places.

Thanks are also due to the City and Guilds of London Institute, Union of Lancashire and Cheshire Institutes and Bolton Institute of Technology for permission to use questions from their past papers on Instrument Technology. These questions have been marked C.G.L.I., U.L.C.I. and B.I.T. respectively.

We thank the British Standards Institution for permission to use information contained in their Temperature Specifications.

The Measurement of Temperature

Temperature is a measure of heat intensity or 'hotness'; it is not heat quantity. We are concerned with two temperature scales. On the ordinary Celsius (centigrade) scale, 0°C and 100°C are defined to be the temperatures of melting ice and boiling water respectively. On this scale the absolute zero of temperature is −273·15°C. For many theoretical purposes, it is convenient to use a scale which starts from absolute zero. This is called the Kelvin scale. Its degree is the same size as that of the Celsius scale, so that to convert from Celsius we need to add 273·15. The ice point is 273·15K.

International Temperature Scale

The International Practical Temperature Scale assigns numerical values to six basic fixed points and thirteen secondary points, as shown. At least two of these points are required to scale a thermometer.

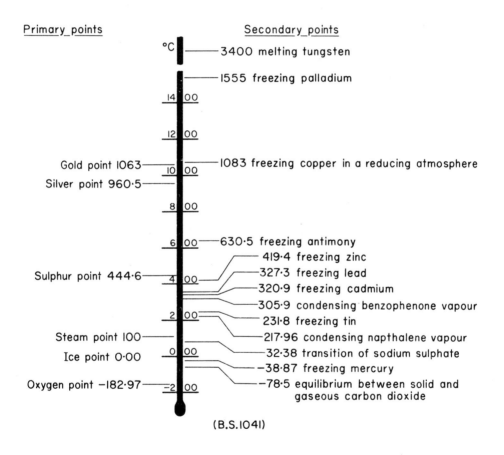

Fig. 1. International Practical Temperature Scale

PART 1 — Change of state

Certain refractory materials when heated start to soften at precise temperatures and can be arranged to indicate various temperature levels. This method of temperature measurement is used considerably in the ceramics industry.

'Seger cones' work on this principle. A range of cones about 5 cm high and of standard shape are made up of mixtures of refractory materials so that each one collapses at a sharply defined temperature.

Temperature-sensitive Paints and Crayons

A chemical change may be utilized as a temperature indicator. The indicator commonly takes the form of a paint or a crayon. Several manufacturers produce a range of papers or crayons whose colours change sharply at a given temperature. The change can take place within a fraction of a second at the indicating temperature, and an accuracy of $\pm 1\%$ is claimed. A common method is to arrange a set of papers in a series with, say, 5 or 10-degree intervals, the set being chosen within a particular range.

This method is useful for difficult situations such as, say, measuring the temperature of the end of a soldering iron.

A limitation is that the changes are irreversible, so that the papers can only be used once. They are useful for spot readings.

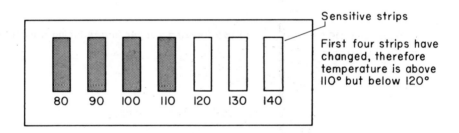

Fig. 2. Temperature-indicating papers

Expansion-type thermometers   PART 2

The operating principle of this type of thermometer is that a solid, a liquid or a gas will expand when heated, the amount of expansion being an indication of the temperature rise.

2.1. Bimetal thermometers

Different metals expand by different amounts if they are subjected to the same temperature rise. If two metals are rigidly fixed together, differential expansion takes place when the metals are heated, causing the composite bar to bend. In one form of bimetal thermometer the element is made of thin bimetal strip wound into a helix. When heated the strip will unwind by an amount proportional to the temperature rise. By making the strip very thin, the response to temperature change is fairly rapid.

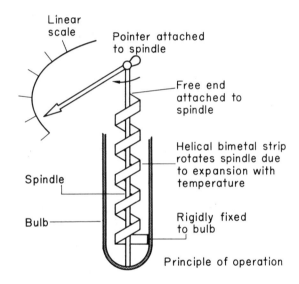

Fig. 3. Bimetal thermometer

The deflection/temperature relationship for many metal combinations is linear over a particular range only, and the materials must be chosen with care.

The industrial range of this instrument is about 0 to 400°C, with an accuracy of ±5% of scale range, though it is possible to produce an accuracy of ±1%. The instrument is rugged, direct-reading, capable of being used under conditions of vibration, and is relatively cheap.

Its range of operation is small and it cannot be used for remote indication or control. The materials used will depend on the particular application, with Invar being commonly used for the low-expansion side, and brass or silicon bronze for the high-expansion side. The full-scale temperature must never be exceeded, as there may be a danger of the metals changing their characteristics at too high a temperature.

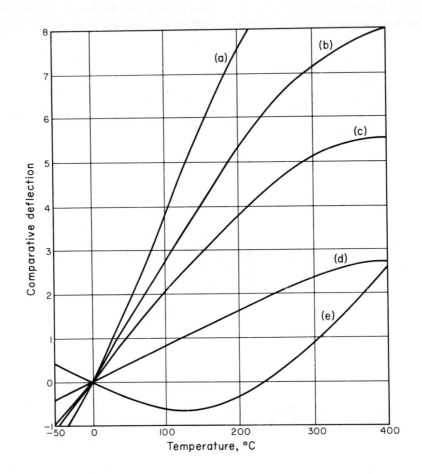

Fig. 4. Comparative deflection/temperature curves for some commercial bimetals

Note: Most are linear over only a small range.
(a) High-expansion side Mg, Cu, Ni
    Low-expansion side Ni, Fe
(b) & (c) Similar materials but proportions vary
(d) High-expansion side pure nickel
    Low-expansion side Cu, Fe
(e) Both sides basically Ni, Fe, Mo, but different proportions

2.2. Bimetal thermostat

The bimetal strip is commonly used for thermostats. Bending takes place when heated, curving away from the metal with the greater coefficient of expansion and breaking the contacts.

As the process cools down, the strip will bend back and close the contacts of the switch.

Assume both metals are of the same thickness, one is Invar with zero coefficient of expansion, the strip is straight at 0°C, and the expansion is measured along the centre lines of the strips.

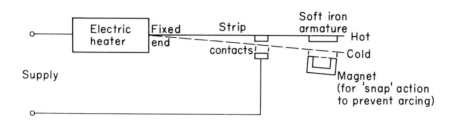

Fig. 5. Bimetal thermostat

Note: we are assuming that the expansion of the metals is linear over the range of temperature in the example.

Length of b for a temperature change t is given by $l_t = l_0(1+\alpha t)$.

Length of a for a temperature change t is given by l_0.

$$\frac{\text{Length of axis of } b}{\text{Length of axis of } a} = \frac{l_0(1+\alpha t)}{l_0} = 1+\alpha t$$

But from the geometry of the figure

$$\frac{\text{length of axis of } b}{\text{length of axis of } a} = \frac{(r+d)\theta}{r\theta} = \frac{r+d}{r}$$

so
$$1+\alpha t = \frac{r+d}{r}$$
$$r + r\alpha t = r + d$$
$$r\alpha t = d$$
$$r = \frac{d}{\alpha t} \quad \ldots\ldots\ldots\ldots (A)$$

From this equation the least radius of curvature, hence the greatest curling, takes place when the strips are very thin.

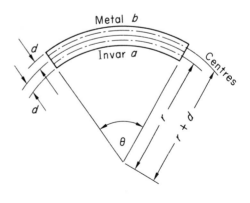

Example.—A bimetal strip is made from Invar and steel strips, each 0·5 mm thick and 50 mm long. If the strip is straight at 0°C and one end is fixed, find the amount of sag at the free end when the temperature of the strip is raised to 50°C. Take α for Invar as being negligible, α for steel as 0·000 013.

From equation (A)

$$r = \frac{d}{\alpha t} = \frac{0\cdot 5}{0\cdot 000\,013 \times 50}$$
$$= \frac{0\cdot 5}{0\cdot 000\,65}$$
$$= 770 \text{ mm}$$
$$PR = QR = 770 \text{ mm}$$
$$\theta = \frac{50}{770} = 0\cdot 0649 \text{ rad} = 3°\,43'$$
$$RS = 770 \cos 3°\,43'$$
$$= 770 \times 0\cdot 9979$$
$$\therefore \text{sag} = PS = PR - RS$$
$$= 700\,(1 - \cdot 9979)$$
$$= 700 \times \cdot 0021$$
$$= 1\cdot 47 \text{ mm}$$

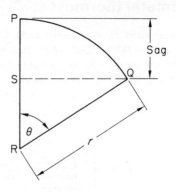

2.3. Liquid-in-glass thermometers

Liquid-in-glass thermometers are widely used and are available in many different forms. They are cheap, reliable and capable of high accuracy. Mercury is the commonest filling liquid though others are also used.

When a liquid is heated it will expand volumetrically. The container will also expand, though the amount will be small and may often be ignored. If the required accuracy is such that the expansion of the container must be taken into account, then the apparent coefficient of expansion—that is the difference of the coefficients of expansion of the liquid and the container—must be considered.

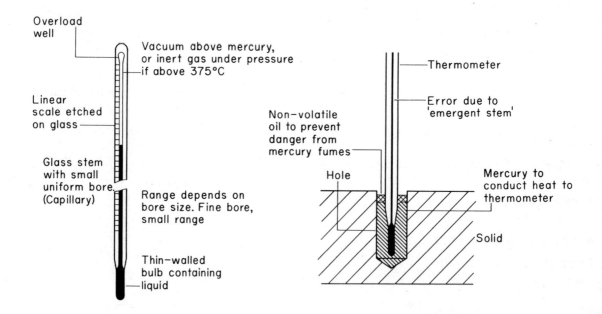

Fig. 6. Typical liquid-in-glass thermometer

Fig. 7. Method of measuring the temperature of a solid

Example.—A mercury-in-glass thermometer contains 1 cm³ of mercury at 0°C. What will be the volume of mercury in the capillary if the bulb and capillary are raised through 40°C? Coefficient of cubical expansion of glass is 0·000 026 and the real coefficient of expansion of mercury is 0·000 18?

Apparent coefficient
$$\begin{aligned}
&= \text{real coefficient} - \text{coefficient of glass} \\
&= 0{\cdot}000\ 18 - 0{\cdot}000\ 026 \\
&= 0{\cdot}000\ 154 \\
V_t &= V_0(1+\gamma t) \\
&= 1(1+0{\cdot}000\ 154 \times 40) \\
&= 1+0{\cdot}006\ 16 \\
&= 1{\cdot}006\ 16\ \text{cm}^3
\end{aligned}$$
∴ volume in capillary = 0·006 16 cm³

Because glass is a bad conductor of heat, the speed of response may be slow, and hence the thermometer must be allowed to reach thermal equilibrium with its surroundings before a measurement is taken. The temperature of a solid, a liquid, or a gas may be measured; with a solid, one method is shown in Figure 7. Mean temperatures are usually indicated in a liquid, efficient stirring being necessary. Problems exist, however, in the measurement of gas temperatures. The thermometers should be fitted with a radiation shield to prevent low readings. Wherever possible the thermometer should be calibrated, used, and standardized to a specified depth of immersion or 'totally immersed' up to the reading.

Glass thermometers may be classified under two main groups, group 1 (N.P.L. standard) being of a higher grade than group 2 (commercial grade). Group 1 instruments are used for calibration purposes or similar work. Group 2 thermometers are for general industrial routine indication; the main requirements are that the thermometers should give consistent readings with reasonable accuracy.

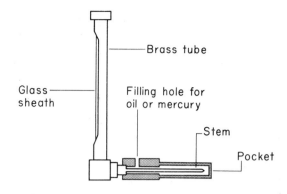

Fig. 8. Brass-cased thermometer with 90° bend

The type of glass used has a big effect on the accuracy. B.S. 1041 specifies a list of glasses approved by the National Physical Laboratory for thermometer use. Careful annealing is essential before calibrating. The same specification shows the minimum accuracy of tests carried out by the N.P.L. for the issue of 'Class A' certificates. All certificated thermometers bear the N.P.L. monogram.

Errors such as zero depression, due to the glass bulb not immediately returning to its original volume after being at a high temperature, and ageing sometimes occur due to faults in the manufacture. Appreciable errors may develop with high-temperature thermometers after long exposure to high temperatures, and may be reduced by using the thermometer only long enough to obtain the required reading.

If subjected to high pressure, as when immersed in a dense liquid, the reading will be high due to decrease in the volume of the bulb.

A range of −200 to +500°C can be achieved by using a variety of liquids. Remote indication or recording is impracticable.

2.3.1. Liquids other than Mercury

Liquids other than mercury are used either where the effects of spilled mercury from a broken thermometer are very undesirable, or where restricted by the temperature range.

2.3.2. Maximum and Minimum Thermometers

Small modifications may be incorporated to produce either a maximum or a minimum thermometer. In the maximum thermometer a rise of temperature causes a large expansion force which easily pushes the mercury through the constriction. Fall of temperature leaves only the thin thread of mercury above the constriction to force it down, hence it will remain at the highest temperature until shaken. This arrangement is used in the clinical thermometer.

In the minimum thermometer surface tension of the alcohol drags an index down with it. Because the viscosity of the alcohol is low, it is unable to move the index with rising temperature and hence the index will indicate the lowest temperature. To reset, the thermometer is tilted until the index is again in contact with the alcohol meniscus.

2.3.3. Calibration

Mercury thermometers are in general tested rather than calibrated. The process of manufacture is of such accuracy that the scale can be impressed on the glass before filling and manufacture completed to the required accuracy. If it is necessary to calibrate a thermometer, it is generally necessary to make a correction table rather than re-mark the scale itself.

B.S. 1041 recommends re-testing at intervals not exceeding five years, and checking the zeros of high-precision instruments at more frequent intervals.

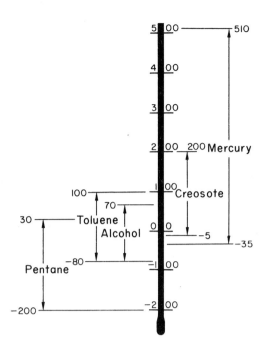

Fig. 9. Range of filling liquids

Fig. 10. Maximum thermometer

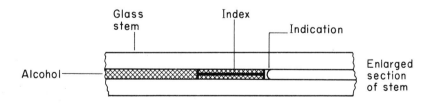

Fig. 11. Minimum thermometer

2.4. Liquid-in-metal thermometers

A liquid-in-metal system consists of a bulb connected to a Bourdon, or some other suitable pressure-actuated element, by a capillary tube. The complete system is sealed under pressure. Expansion of the liquid, due to temperature increase, will deflect the measuring element by an amount proportional to the temperature rise. Spiral Bourdon tubes are often used since a greater deflection is obtained for the same temperature rise. Although a pressure element is used for indication, the deflection is due to change of volume rather than change of pressure.

2.4.1. Bulbs

The bulb is the sensitive part of the system. Good response speed is achieved by means of a large surface to volume ratio, by making the mass of metal as small as possible and by avoiding the use of polished surfaces which would restrict heat transfer. The type of filling liquid will also have an effect on the response speed. Usually the bulb is made long and narrow.

Bulb and capillary materials are mostly mild or stainless steel, copper or Monel. The choice will be determined by the filling liquid and the particular installation.

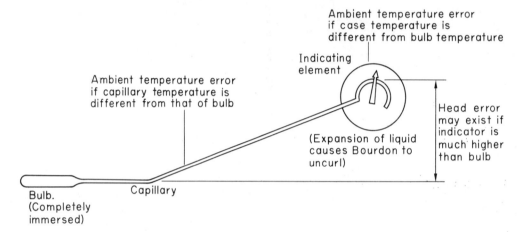

Fig. 12. Principle of the liquid-in-metal thermometer showing some possible errors

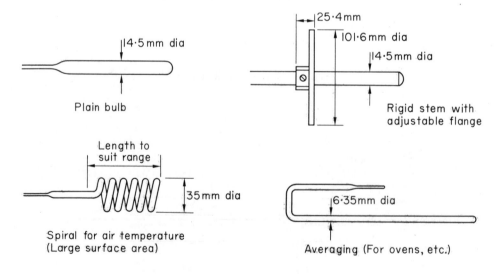

Fig. 13. Thermometer bulbs

2.4.2. Capillaries

Capillaries are made from seamless cold-drawn tubes; stainless steel is commonly used with mercury. Internal diameters vary between about 0·1 to 0·5 mm, with lengths up to 60 metres (about 200 ft).

2.4.3. Ambient-temperature Errors

Ambient-temperature effects may cause errors in both the capillary and the indicating element. The amount of error will be given by: error = $(v/V) \times t$, where v, V and t are volume of capillary, volume of bulb and ambient temperature change respectively.

Example.—A mercury-in-steel thermometer bulb has a volume of 3000 mm³ and a capillary volume of 250 mm³. What will be the error in the reading if the ambient temperature changes by +15°C? If the instrument has a range of 100 to 600°C, what is the percentage range error?

$$\text{Error} = \frac{250}{3000} \times 15 = 1\cdot25°C$$

$$\begin{aligned}\text{Range error} &= \frac{1\cdot25}{600-100} \times 100 \\ &= \frac{1\cdot25}{500} \times 100 \\ &= 0\cdot25\%\end{aligned}$$

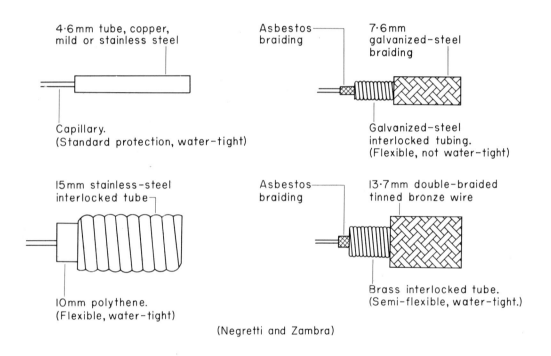

(Negretti and Zambra)

Fig. 14. Various types of capillary tube

2.4.4. Ambient-temperature Compensation

Errors introduced by the capillary and indicating element may be compensated by fitting an additional capillary and a differential Bourdon element.

The Negretti and Zambra 'Mersteel' system uses compensating links. Figure 16 shows the principle of operation. The number of links required depends on the length of the tubing.

2.4.5. Case Compensation

To compensate for the indicating element alone, a bimetal strip is connected between the end of the Bourdon tube and the pen or pointer. As the ambient temperature varies, the strip will deflect in such a way and by such an amount as to cancel out the error.

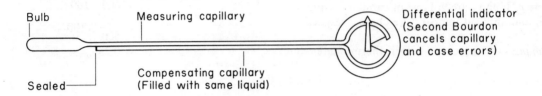

Fig. 15. Use of compensating capillary

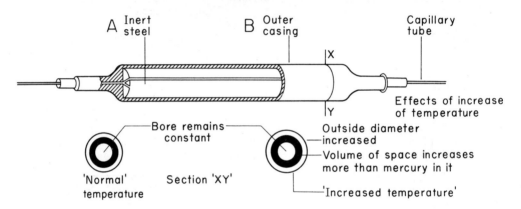

Fig. 16. Negretti and Zambra compensating link

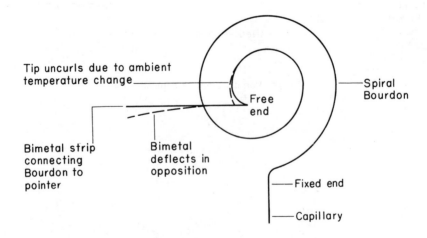

Fig. 17. Bimetal temperature compensator

2.4.6. Other Errors

In addition to the errors mentioned, other possible errors are as follows:

Hysteresis is the difference between the 'up' and the 'down' readings, and may be overcome by careful filling with mercury and selection of the Bourdon tube.

Ageing is caused by being calibrated before the metal of the bulb has really settled down. This can be overcome by ageing the thermometer before calibration.

Dip effect: If the temperature is increased suddenly, the bulb will momentarily expand faster than the mercury. This will cause mercury to flow into the bulb, apparently indicating a sudden drop in temperature.

2.4.7. Other Filling Liquids

Although mercury and steel are the most common combinations, other metals and filling liquids are used, the temperature range of the liquids being shown in Figure 18. The choice of metal will be determined by the type of filling liquid and the nature of the fluid whose temperature is being measured. It may, for example, be corrosive or abrasive.

2.4.8. Refilling

It is not always possible to refill a liquid-in-metal thermometer, but if there is a screw coupling between the case and the capillary, and great care is taken, refilling may be attempted.

After first evacuating, both parts of the system are filled with liquid under a pressure of about 7 MN/m^2 (70 bar) (1000 lbf/in^2 approx.) to ensure that there are no air locks. The pressure is then released, and both parts are screwed together, as shown, taking care that no air enters the system. It may be necessary to loosen the coupling and adjust the amount of liquid to obtain the correct filling; this will be indicated by obtaining the correct reading at maximum temperature. After checking all the main points, the thermometer should be aged by maintaining it at the maximum temperature for several days.

Linkage errors are likely to occur just as for a pressure gauge, and must be corrected in a similar manner.

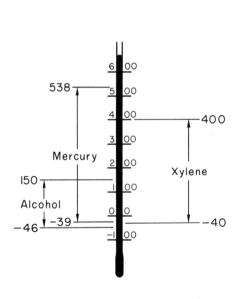

Fig. 18. Filling liquids

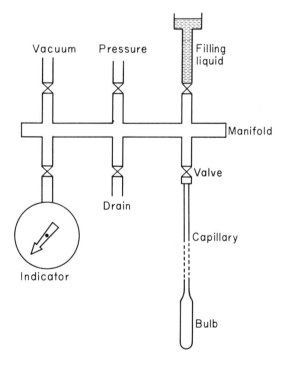

Fig. 19. Method of refilling

2.4.9. Range

Liquid-in-metal thermometers may be used for measuring the temperature of solids, liquids or gases. Typical industrial ranges are from 0 to 300°C with an accuracy of $\pm 0.5\%$ or 0 to 600°C with an accuracy of $\pm 1\%$ of scale range, for mercury-in-steel. These thermometers are suitable for remote indication up to 60 metres (about 200 ft).

2.4.10. Pockets and Sheaths

If the temperature medium is under pressure or is corrosive or abrasive, the use of a pocket will protect the bulb. Fitting and removal of the thermometer is also simplified. Sensitivity will be decreased and may cause low readings if the external parts of the connection, together with the adjacent radiating surface, are not lagged. It is also essential to have good thermal contact between the pocket and bulb; this may be achieved by using a corrugated metal insert or filling the pocket with a suitable liquid.

Sheaths are often fitted to bulbs which are exposed to atmospheres such as salt baths or furnaces. The materials used for sheaths and pockets will depend on the temperature medium and may be either metal or refractory.

2.5. Temperature regulators

Although automatic control systems as such are beyond the scope of this book, many types of comparatively simple regulators operating on the principle of liquid expansion are available. One such regulator is shown in Figure 23.

Expansion of the oil with rising temperature moves the valve towards the closed position. When the temperature falls the oil contracts and the return spring opens the valve once more.

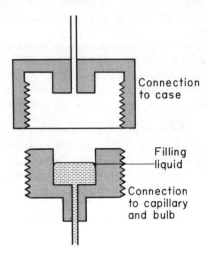

Fig. 20. Method of connecting

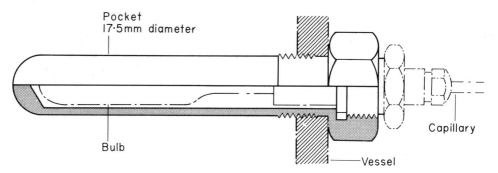

Fig. 21. Thermometer pocket

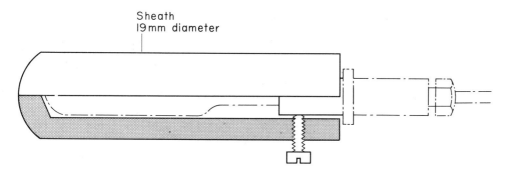

Fig. 22. Thermometer sheath

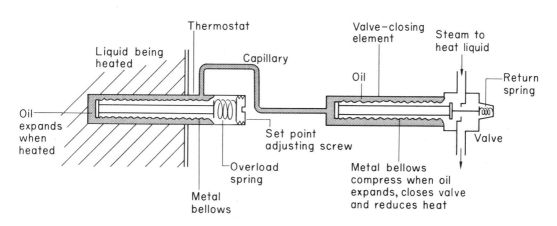

Fig. 23. 'Sarco' temperature regulator

PART 3 — Pressure-type thermometers

A gas or vapour is substituted for a liquid in a sealed system which permits negligible change of volume. As a result, an increase of temperature causes an increase of *pressure* in the gas which is measured directly by a pressure gauge.

3.1. Gas-filled thermometers

The gas thermometer consists of a bulb, capillary and some form of pressure-measuring element, usually a Bourdon tube. The gas is usually nitrogen, and the working pressure is relatively high (about 10 bar). The operation of the thermometer is dependent on the laws of perfect gases, which may be expressed as $P_2/P_1 = T_2/T_1$ where the volume is kept constant (a reasonable assumption in this application). P_1, T_1, P_2, T_2 are the initial and final pressures and temperatures respectively, and are in absolute values. Deviation from the ideal gas law is negligible if used only for general industrial purposes.

Rearranging: $\dfrac{P_2}{P_1} - \dfrac{P_1}{P_1} = \dfrac{T_2}{T_1} - \dfrac{T_1}{T_1}$

or $\dfrac{P_2 - P_1}{P_1} = \dfrac{T_2 - T_1}{T_1}$

$P_2 - P_1$ represents the pressure change in the system corresponding to a temperature change of $T_2 - T_1$.

Example.—Hydrogen at an absolute pressure of 15 bars and at 50°C, the minimum temperature reading, is introduced into a gas thermometer system. What will be the pressure of the gas when the maximum temperature, 200°C, is indicated on the thermometer?

$$\frac{P_2 - 15}{15} = \frac{200 - 50}{323}$$

$$\frac{P_2 - 15}{15} = \frac{150}{323}$$

$$P_2 - 15 = 15 \times 0\cdot 465$$

$$P_2 = 6\cdot 98 + 15$$

$$= 21\cdot 98 \text{ bar}$$

Errors will be introduced due to ambient-temperature changes, and may be compensated as for the mercury-in-steel type discussed. Unfortunately, the capillary error is a function of, and increases with, the temperature of the bulb. Hence, compensation can only be completely effective at one particular temperature. It is usual to make the ratio

$$\frac{\text{bulb volume}}{\text{capillary} + \text{gauge volume}}$$

high in order to keep ambient errors to a minimum.

It is clear, from the equation, that the scale is linear. Speed of response is high and the cost is relatively low. A typical range for industrial applications is 0 to 550°C, with an accuracy of ±1% of scale range, with indication up to 60 metres (about 200 ft).

Disadvantages are large bulb size and small differential range, that is the difference between maximum and minimum temperatures.

3.1.1. Refilling

The following procedure may be adopted, if the method of manufacture allows:

The system should first be heated to prevent any moisture from introducing errors due to vapour pressure. The air which will be present will not cause any error. The connecting capillary must be soldered to the filling system and nitrogen allowed to enter until the pointer reads the first scale point with the bulb heated to the appropriate temperature. Half and full scale points may now be checked. A zero error may be corrected by either removing or adding an appropriate amount of nitrogen.

When the readings are correct, the capillary end is sealed by pinching and soldering. Check for any leakage. Ageing should then be carried out as for the mercury thermometer.

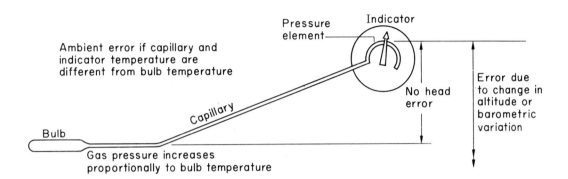

Fig. 24. Gas-filled thermometer system

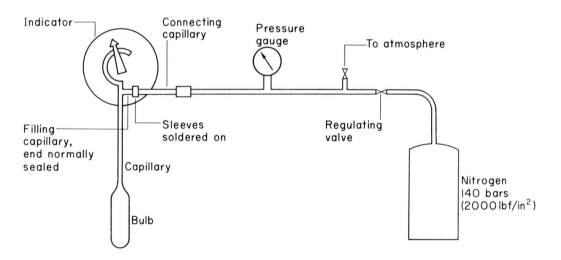

Fig. 25. Refilling of gas thermometer

3.2. Vapour pressure thermometers

A vapour may be defined as a gas which can be liquefied by compressing it without having to cool it as well. The temperature to which a gas must first be reduced before it will liquefy is called its *critical temperature*. Consider a portion of liquid enclosed in a vessel along with its own vapour (no air present). A vapour under these conditions is described as *saturated*. At a given temperature molecules of the liquid will evaporate into the vapour, and at the same time molecules of the vapour will condense back into the liquid. Given a little time, the pressure will adjust itself to a value at which the two processes are in equilibrium. This is called the *saturation vapour pressure*, or just the *vapour pressure*, and depends only on the temperature and the nature of the liquid. This forms a useful method of temperature measurement and one that is not subject to the errors due to ambient temperature that affect most methods. Different vapours exert different saturation vapour pressures at any given temperature.

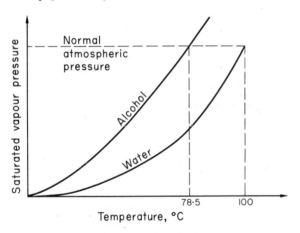

Fig. 26. Variation of S.V.P. with temperature

Vapour pressure systems follow a similar construction to liquid and gas thermometers, except that the bulb is *partially* filled with a volatile liquid suitable for the temperature range. There must be enough fluid to ensure that the liquid and vapour interface will always be in the bulb.

The capillary and Bourdon will be filled with either liquid or vapour depending on the conditions shown in Figure 27.

3.2.1. Errors

Head errors can exist due to a column of liquid in the capillary as shown in Figure 27. Corresponding but reversed effects occur when the gauge is mounted below the bulb. There may be an error due to variation of barometric pressure.

When the bulb temperature changes from slightly above to slightly below the gauge and capillary temperature, the condensed vapour originally in the gauge and capillary will distil in the bulb, leaving the gauge and capillary void of liquid but filled with superheated vapour. This transfer of fluid will require a definite time to take place. Similarly when changing in the opposite direction. This is known as *cross ambient effect* and results in sluggishness. This type of thermometer is not recommended where temperatures are likely to fluctuate in this manner.

If a bulb is suddenly cooled from a temperature above to a temperature below that of the gauge and capillary, momentary superheating of the liquid in the gauge and tubing causes the pointer to kick upwards. This should not be taken as a defect in the instrument.

3.2.2. Overcoming Cross Ambient Effects

It is possible to overcome cross ambient effects by adding a second non-volatile liquid into the bulb which separates the liquid from the vapour and acts as the pressure-transmitting medium.

3.2.3. Refilling

It may be possible to refill the instrument using the equipment shown in Figure 29.

The instrument should first be checked with a gas at a pressure equal to the full-scale pressure that would be produced by the vapour at full-scale temperature. After this the system should be evacuated for two or three hours to remove all air. With the bulb at about $-20°C$ a calculated amount of the correct filling liquid should be introduced. Checks may then be taken at full, half and bottom scale readings.

A range of -20 to $350°C$ is typical, with lengths up to 60 metres (about 200 ft), with an accuracy of $\pm 1.5\%$. They are highly sensitive within a narrow temperature range (depending on the filling fluid). The highest point must be well below the critical temperature. The scales are non-linear.

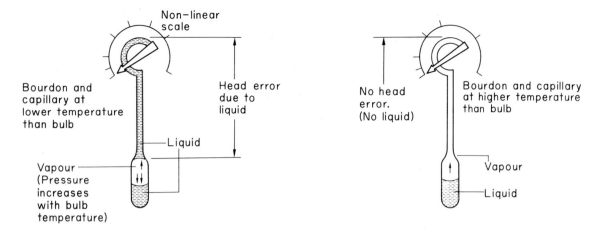

Fig. 27. Vapour pressure thermometers

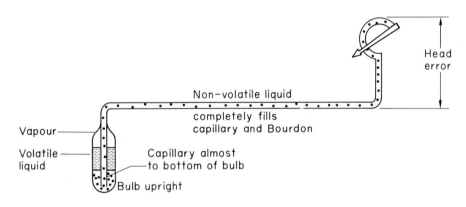

Fig. 28. Two-liquid thermometer

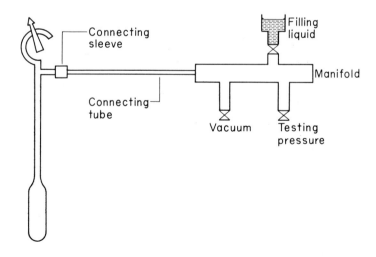

Fig. 29. Refilling vapour pressure thermometers

Table 1. Properties of some filling liquids

Liquid	Boiling-point, °C	Critical temperature, °C	Typical range, °C
Methyl chloride	−23·7	143·1	0–50
Butane	0·6	153·0	20–80
Ethyl chloride	12·2	187·2	30–100
Diethyl ether	34·5	193·8	60–160
Ethyl alcohol	78·5	243·1	100–180
Water	100·0	374·0	120–220
Toluene	110·5	320·6	150–250

3.2.4. Remote Transmission

If the indicating point is beyond about 60 metres some form of transmitter must be introduced. A typical pneumatic transmitter is shown in Figure 30.

Temperature increase at the bulb A tends to move the Bourdon tube B in the direction indicated, lifting the valve-operating frame C and allowing the valve D, fed with an air supply of 1·4 bar (20 lbf/in²) at F, to move so that its output pressure increases. This increase in pressure is fed back through a damping restriction on to diaphragm E, thus restraining the Bourdon tube and restoring the state of balance. Since the torque exerted by the Bourdon is proportional to the effect of the feedback pressure, it follows that this pressure, fed to the line as the transmitter output, is directly proportional to the bulb temperature.

3.2.5. General

The type of expansion thermometer used for a particular temperature-measuring installation will depend on several factors such as: speed of response, whether a linear scale is required, whether good discrimination is required at a particular temperature, etc. Figure 32 shows the speed of response of several thermometers.

Mechanical thermometers are inherently more robust than electrical thermometers as the bulb material is never critical to the working of the device. They are completely self-contained, and no power supplies or ancillary equipment are required. Most forms can be adapted to feed into control systems but less easily than electrical thermometers.

3.2.6. Pipe Runs

If the capillary is long, care must be taken to ensure that other piping, etc., is not added at a later date which will prevent the removal of the instrument, should this become necessary. Tubing must be secured by clips. Any surplus length should be coiled and fixed to the back of the panel out of the way. Sharp bends should be avoided, and the tubing should not be embedded in concrete. No hot pipes should come in contact with the capillary, particularly if compensation is not provided.

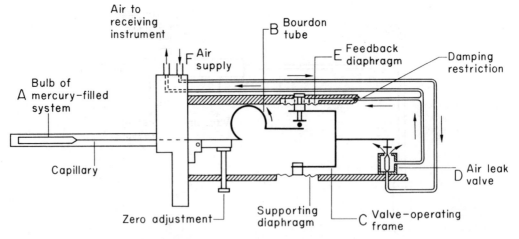

Fig. 30. Negretti and Zambra pneumatic temperature transmitter

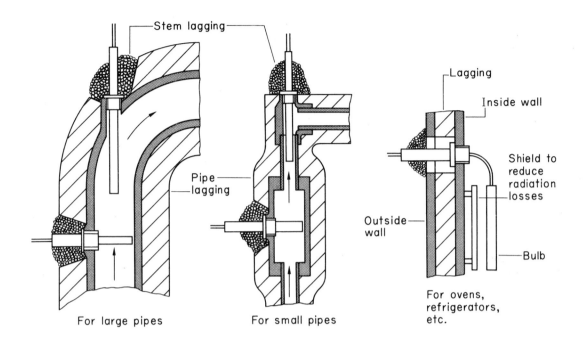

**Fig. 31. Typical bulb installations
(Budenberg Gauge Co.)**

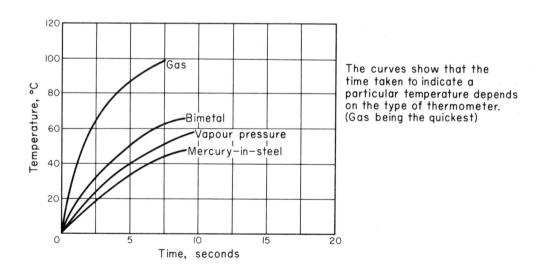

The curves show that the time taken to indicate a particular temperature depends on the type of thermometer. (Gas being the quickest)

Fig. 32. Response curves for various thermometers

Table 2. Summary of mechanical thermometers

Type	Common Combinations	Relative Cost	Range °C	Response Speed	Remarks
Paints and crayons		Cheap	30/250		Can only be used once
Liquid in glass	Mercury Alcohol	Cheap	−200/+500	Slow	Cannot record
Liquid in metal	Mercury in steel	Medium	0/600	Slow	Can record and control
Bimetal	Nickel, iron	Cheap	0/400	Medium	No remote indicating
Gas expansion	Nitrogen	Medium	0/550	Fast	Linear scale
Vapour pressure		Medium	−20/+350	Medium	Non-linear scale

City and Guilds Examination Questions

1. A vapour pressure thermometer is to be installed to measure the temperature of an oil bath in the range 60°C to 100°C, the indicator being situated fifteen feet below the temperature-sensing element.
   (*i*) Explain the principle of operation of the instrument and state the errors to which it is liable.
   (*ii*) Suggest a suitable filling liquid and sketch a typical curve showing the relationship between its vapour pressure and temperature.
   (*iii*) Describe briefly the procedure for refilling the system following a fracture of the capillary.
   (310/3 1967)

2. Describe, with diagrams, the principle of operation of *one* of the following thermometers, giving the scale shape associated with it:
   (*i*) mercury in steel;
   (*ii*) gas-filled;
   (*iii*) vapour pressure.
Explain the effect (if any) on the performance of the instrument described of:
   (*a*) large changes in temperature of the capillary;
   (*b*) position of the bulb relative to the indicator;
   (*c*) large barometric pressure changes.
   (79/2 1963–4)

3. (*a*) Describe in detail, with appropriate sketches, the principle of operation of a mercury-in-steel industrial thermometer.
   (*b*) Explain how ambient temperature changes can affect the performance of the instrument and describe *two* different compensation systems which can be built into the instrument.
   (79/2 1965–6)

4. (*a*) Name *three* different designs of thermal expansion element used in industrial measurement and control.
   (*b*) Sketch the basic construction of an industrial bimetal thermometer using *one* of the above elements.
   (*c*) Give one industrial application of the type of thermometer sketched.
   (*d*) Sketch a graph of deflection against temperature showing typical deflection characteristics of a bimetal strip.
   (*e*) A rod-type thermostat consists of elements having thermal expansion coefficients of 0·000 026 and 0·000 001. What length of element would be required to cover a temperature range of 0 to 100°C with one turn of a setting screw which has a lead of 1 mm?
   (400 mm) (310–1–03 1969)

Electrical thermometers — PART 4

Electrical thermometers consist of two groups; those whose resistance changes with temperature (resistance thermometers) and those where a change of temperature produces a change of e.m.f. (thermocouples). Both types are widely used in industry.

4.1. Resistance thermometers

Resistance elements are normally manufactured so that they have a nominal ohmic value at 0°C. Their resistance is then checked at this temperature and a 'make-up' resistance is connected in series with the element so that the thermometer has a definite resistance of, say, 100 ohms at 0°C. This make-up resistance must not be confused with the ballast resistance used in multipoint installations.

The resistance of such an element does not vary linearly with temperature, so that when it is calibrated more than two points must be taken. The change of resistance of the element as its temperature is raised through 0 to 100°C is known as the *fundamental interval*, and within a working temperature range of 0 to 660°C an adequate approximation of the resistance/temperature relationship is given by $R_t = R_0(1 + At + Bt^2)$, where t is the temperature rise of the element above 0°C, and R_t is the resistance at that temperature. The constants R_0, A and B are determined at the ice, steam and sulphur points respectively. For a platinum thermometer the ratio R_t/R_0 must not be less than 1·390 for $t = 100°C$ and 2·645 for $t = 444·6°C$.

4.1.1. Callendar-Van Dusen Equation

The relation of resistance to temperature between $-182·97°C$ and $630·5°C$ is defined by the following Callendar—Van Dusen equation:

$$\frac{R_t}{R_0} = 1 + \alpha \left[t - \gamma \left(\frac{t}{100} - 1 \right) \left(\frac{t}{100} \right) - \beta \left(\frac{t}{100} - 1 \right) \left(\frac{t}{100} \right)^3 \right]$$

where R_t is the element resistance at $t°C$, R_0 is the resistance at 0°C, and α, β and γ are characteristic constants. Typical values are $\gamma = 1·49$; $\beta = 0·11$ for negative t and 0 for positive t; $\alpha = \dfrac{R_{100} - R_0}{100 R_0}$.

Example.—The relationship between resistance and temperature of a platinum resistance thermometer between the ice point and 660°C is given by $R_t = R_0(1 + At + Bt^2)$. If $A = 39·08 \times 10^{-4}$ and $B = -0·58 \times 10^{-6}$ and $R_0 = 100$ ohms, determine the change in resistance of the thermometer if its element is raised from 0° to 100°C.

$$R_t = R_0(1 + At + Bt^2)$$
$$R_{100} = 100[1 + 39·08 \times 10^{-4} \times 100 + (-0·58 \times 10^{-6}) \times 100^2]$$
$$= 100(1 + 0·3908 - 0·0058)$$
$$= 100(1·3908 - 0·0058)$$
$$= 100(1·385)$$
$$= 138·5$$
$$\therefore \text{Change} = 138·5 - 100$$
$$= 38·5 \text{ ohms}$$

Example.—From the Callendar-Van Dusen relationship determine the ratio R_t/R_0 when the temperature is raised from 0° to 100°C. $\gamma = 1·49$, $\beta = 0$ and $R_{100} = 139$ ohms.

$$\alpha = \frac{139 - 100}{100 \times 100} = \frac{39}{10000} = 0·0039$$
$$R_t/R_0 = 1 + 0·0039 \left[100 - 1·49 \left(\tfrac{100}{100} - 1 \right) \left(\tfrac{100}{100} \right) - 0 \right]$$
$$= 1 + 0·0039(100 - 0)$$
$$= 1 + 0·39$$
$$= 1·39$$

4.1.2. Resistance Thermometer Elements

Resistance elements may be classed as wire-wound, photo-etched and thermistor types.

4.1.3. Wire-wound Elements

The dimensions of wire-wound elements are determined by winding resistance, time constant, self heating, etc. The temperature range is -240 to $600°C$, with an accuracy of $0·75\%$.

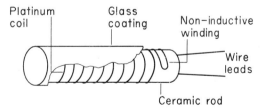

Rosemount Engineering Co.
High-temperature sensor

Fig. 33. Example of wire element

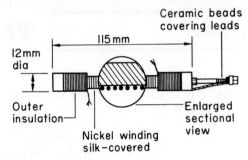

Fig. 33. Example of wire element

The nominal resistance of the sensor made by Budd Instruments is 50 ohms at 24°C, about 100 ohms at 190°C. Percentage change of resistance with temperature is shown on the graph.

Because of its small mass a very rapid response to temperature change is inherent. Wire leads should be soldered with a miniature pencil-type iron in order not to damage the sensor.

4.1.4. Time Constant

Change of resistance due to a step change in temperature is exponential. The time constant is expressed as the time taken for the resistance to change to 63·2% of the final value. The response time depends on a number of factors, such as size of element, if sheathed, etc.

4.1.5. Self Heating

For precise measurement an allowance for increase in indicated temperature due to the I^2R heating of the element must be made. The power dissipated must normally be kept below 20 mW.

4.1.6. Construction

A typical assembly of a complete unit for measuring between −200°C and 550°C is shown in Figure 34.

4.1.7. Photo-etched Elements

Photo-etched or printed-circuit resistance elements are useful for measuring the surface temperatures of the component to which they are bonded. In appearance they are very similar to strain gauges (described in *Instrumentation: Pressure and Liquid Level*), the conductor being of nickel.

The temperature range is −200 to +300°C, to an accuracy within 0·5°C.

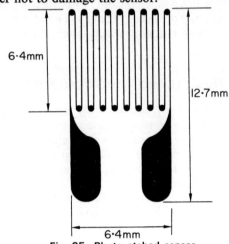

Fig. 35. Photo-etched sensor

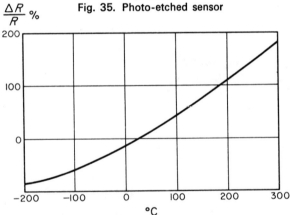

Fig. 36. Variation in resistance with temperature (Budd Instruments)

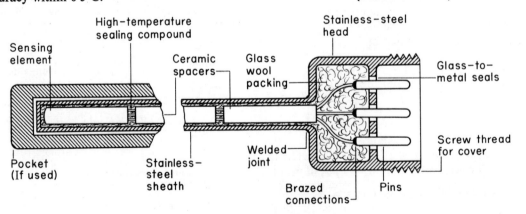

Fig. 34. Resistance thermometer assembly (Rosemount Engineering)

4.1.8. Parasitic Phenomena

A photo-etched sensor will, like the strain gauge, be sensitive to strains. This effect is difficult to allow for as the response to strain is generally far from linear. However, the effect of this compared to temperature change is small and may, in general, be ignored.

4.1.9. Thermistor Elements

Thermistors are a type of semiconductor made from certain metal oxides and formed into various shapes and sizes. They have both positive and negative temperature coefficients of resistance, the negative one being used for temperature measurement. A typical value of coefficient is -4% per °C at 20°C. They are made in three main forms: bead, disc and rod.

Bead types may be directly or indirectly heated. Disc types are cheaper than beads, larger in size and capable of handling higher power, but are slower in response. Rod types generally have a higher mass and are not normally used for temperature measurement. The temperature range, at the moment, extends from well below 0°C to 300°C; disc types are limited to a maximum of about 125°C.

4.1.10. N.T.C. Thermistor Characteristic

The resistance/temperature characteristic of a typical negative temperature coefficient thermistor, as plotted in Figure 39, approximates to an exponential relationship, hence

$$R = Ae^{B/T}$$

where
- R = thermistor resistance (ohms)
- T = thermistor temperature (K)
- B = thermistor 'characteristic temperature' (K), an arbitrary constant not related to any specific value of R or T
- A = a constant.

If R_1 is the resistance at temperature T_0 and R_2 is the resistance at temperature T_2, then

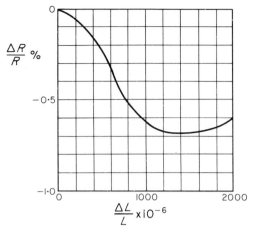

Fig. 37. Effect of strain on resistance

$$R_1 = Ae^{B/T_1} \quad \text{and} \quad R_2 = Ae^{B/T_2}$$
$$\frac{R_1}{R_2} = \frac{e^{B/T_1}}{e^{B/T_2}} = e^{B/T_1} \cdot e^{-(B/T_2)} = e^{B(1/T_1 - 1/T_2)}$$
$$\therefore R_1 = R_2 \, e^{(B/T_1 - B/T_2)} \quad \ldots \ldots \ldots \ldots (i)$$

Typical values for B are between 2500 and 5000K.

Example.—Given that the resistance of a thermistor at 20°C is 500 kilohms and its resistance at 25°C is 390 kilohms, find the characteristic temperature (B).

Using equation (i) and putting the temperature in K,
$$500\,000 = 390\,000 e^{(B/293 - B/298)}$$
Taking logarithms
$$\log 500\,000 = \log 390\,000 + (B/293 - B/298) \log e$$
$$5\cdot6990 = 5\cdot5911 + (B/293 - B/298) \times 0\cdot4343$$
$$0\cdot1079 = \frac{298B - 293B}{293 \times 298} \times 0\cdot4343$$
$$0\cdot1079 = \frac{5B}{293 \times 298} \times 0\cdot4343$$
$$\therefore B = \frac{293 \times 298 \times 0\cdot1079}{5 \times 0\cdot4343}$$
$$= 4338°K$$

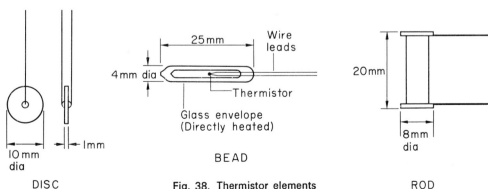

Fig. 38. Thermistor elements

4.1.11. Linearization

Because the R/T characteristic is non-linear, the resulting temperature scale will be non-linear, but if this is undesirable, it is possible to design a circuit to produce a linear characteristic over a limited range.

4.1.12. Advantages and Disadvantages

Thermistors are robust, have a temperature coefficient about ten times that of platinum at room temperature, are physically small, hence have a fast response time, and are available in many resistance ranges. Their cost, however, depends on the closeness of tolerance.

Their main disadvantages are that they are less stable than metal elements, and their characteristics can change with use, though stability will increase with age.

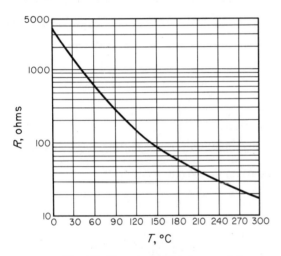

Fig. 39. Thermistor R/T characteristic

4.1.13. Measuring Circuits

The change of resistance of the element is usually measured by some form of Wheatstone bridge and may be used either in the null (balanced) condition, or in the deflection (out-of-balance) condition.

4.1.14. Principle of Operation

If the galvanometer has a centre zero, the resistances are arranged so that the bridge is balanced at a particular temperature. As the bulb temperature varies, so does its resistance, and the bridge is thrown out of balance, the pointer is deflected over the scale, the degree of deflection being an indication of the temperature. Any variation of voltage will affect the circuit sensitivity, and hence R_v is used to correct the voltage in a simple installation.

A rectified and stabilized mains power supply will completely eliminate voltage variation.

Considerable reduction of voltage variation error may be achieved by using a crossed-coil type of indicator rather than the conventional moving-coil galvanometer. The principle of operation is shown in Figure 41 where currents flowing in the coils will set up torques which will oppose each other, the net torque being due to the current flow through each, which is affected equally by any change of supply voltage.

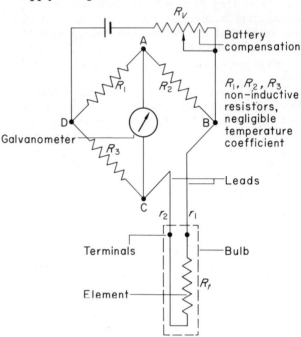

Fig. 40. Two-wire system of connection for the deflection-type circuit

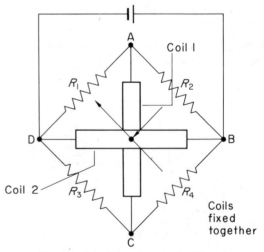

Fig. 41. Principle of crossed-coil meter

4.1.15. Lead Resistance

With the two-wire arrangement the bulb leads r_1 and r_2 are in the same arm of the bridge and any variation of their resistance due to ambient-temperature variation will cause an error.

To keep the lead resistance low, thick copper conductors may be used, though this becomes uneconomical if long distances are involved. B.S. 1041 specifies that the thermometer resistance should be at least 30 times that of the leads in order to reduce ambient errors.

Ambient errors may be reduced by modifying the basic circuit to form three- or four-wire circuits.

4.1.16. Three- and Four-Wire Circuits

The three-wire arrangement is due to Sir William Siemens, and may be considered analogous to the compensating capillary in the vapour-pressure and liquid-in-metal thermometers. Complete compensation, however, is only attained at one particular temperature.

The four-wire arrangement is a later modification due to Callendar. Any change of resistance of the leads due to ambient temperature changes will be cancelled out across the bridge.

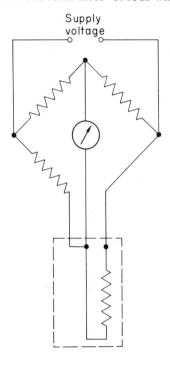

Fig. 42. Three-wire system

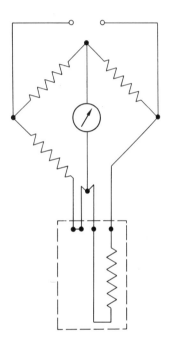

Fig. 43. Four-wire system

Example.—At a certain temperature, when the bridge is balanced, all four resistors have the same value. What will be the resistance of R_t if the power dissipated in it is 20 mW?

If the circuit is redrawn it is clear that R_1 and R_2 are in parallel with R_3 and R_t in series. All resistance values are the same, then the volt drop across R_t will be 1 volt.

$$\text{Power} = W = \frac{V^2}{R} \quad \therefore \quad 0{\cdot}02 = \frac{1}{R}$$
$$0{\cdot}02R = 1$$
$$\therefore \quad R = 50 \text{ ohms}$$

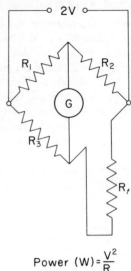

Power $(W) = \dfrac{V^2}{R}$

4.1.17. Multipoint Installations

One indicator may be used to measure the temperature of several different points. Single-pole switches reduce contact resistance, although with double-pole switches there is complete isolation except for the the circuit which is 'on'. Ballast resistors are included to equalize the total circuit resistance for each point.

A continuous method of monitoring a number of temperature points may be carried out by means of an electronic scanner, motor-driven switch or uniselector, which enables sequential switching to take place between the various points.

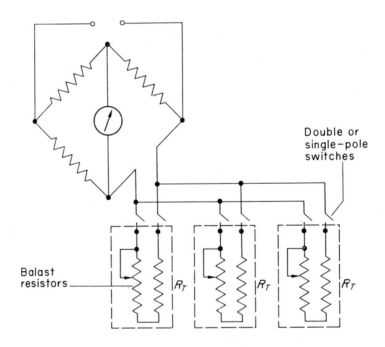

Fig. 44. Multipoint installation

4.1.18. Differential Temperature

Differential temperature may be measured as shown in Figure 45.

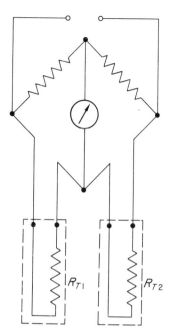

Fig. 45. Differential temperature

4.1.19. Use of Wheatstone Bridge

The Wheatstone bridge is suitable for photo-etched elements, although, if the temperature range is only a few degrees, the use of a strain-gauge indicator is recommended.

The supply voltage may be alternating, and then a simple galvanometer detector will no longer be suitable. Care must also be taken to ensure that errors due to capacitance or inductance will be negligible. The supply voltage must be reasonably high for good sensitivity, but not so high that undue self-heating of the resistors becomes objectionable. For the same reason, dissimilar metals in contact could set up thermoelectric voltages. About 20 volts would be a maximum.

Contact resistance may cause errors if connections are not tight; soldered connections should be used unless connections must be removable. Where switching is used, self-cleaning contacts are advisable.

4.2. Thermocouple thermometers

If two dissimilar pure metal or alloy wires are joined together at each end to form a loop, and a difference of temperature exists between both ends, a difference in junction potentials will be set up resulting in a thermoelectric e.m.f. This is known as the *Seebeck effect*. The magnitude of the e.m.f. will be determined by the particular materials used and the temperature difference between the two ends.

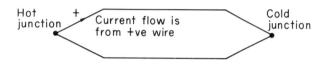

Fig. 46. Simple thermocouple

If one junction is maintained at a constant and known temperature (the reference or cold junction), and the characteristics of the thermocouple are known, the magnitude of the e.m.f. generated will be a measure of the temperature of the other junction (the hot junction). *The e.m.f. generated for any particular two metals at a given temperature will be the same, regardless of the size of the wires, the areas in contact, and the method of joining them together.*

The relationship between temperature and generated e.m.f. is non-linear except over very limited ranges. On the steep part of a curve they may be connected by the approximate empirical laws of the type:

$$e = a(t_1 - t_2) + b(t_1^2 - t_2^2)$$

where e is the generated e.m.f. (volts), t_1 and t_2 are the temperatures of the hot and cold junctions respectively (K), and a and b are constants and depend on the thermocouple material.

4.2.1. General Form of E.M.F./Temperature Curve

If the temperature of the hot junction is continually raised, the e.m.f. will rise to a maximum then begin to fall, and finally will reverse in direction. This is known as *thermoelectric inversion*. The hot-junction temperature for which the e.m.f. is maximum is called the *neutral temperature*. It is important not to use metals whose neutral temperature is within the range to be measured.

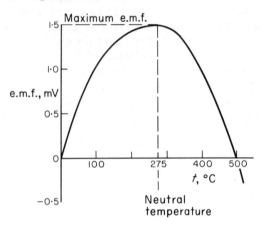

Fig. 47. Thermoelectric e.m.f. for Cu/Fe thermocouple

Example.—A copper/constantan thermocouple has a cold-junction temperature of 27°C and a hot-junction temperature of 127°C. Calculate the e.m.f. generated if the constants a and b are 37.54×10^{-3} mV/°C and 0.045×10^{-3} mV/°C respectively.

$$\begin{aligned}
e &= 37.54 \times 10^{-3}(400-300) \\
&\quad +0.045 \times 10^{-3}(400^2-300^2) \\
&= 37.54 \times 10^{-3} \times 100 \\
&\quad +0.045 \times 10^{-3}(160\,000-90\,000) \\
&= 3.754 + 0.045 \times 10^{-3} \times 70\,000 \\
&= 3.754 + 3.15 \\
&= 3.904 \text{ mV}
\end{aligned}$$

The e.m.f. equation may also be written in the form

$$\log_{10} e = A \log_{10} T + B$$

where e is the e.m.f. (μV), T is the temperature of the hot junction (°C), *the cold junction being held at 0°C*, and A and B are constants.

Example.—Calculate the e.m.f. of a platinum/platinum 10% rhodium thermocouple when the hot junction is at 250°C and the cold junction is at the ice point. Take A and B as 1.19 and 0.52 respectively.

$$\begin{aligned}
\log_{10} e &= 1.19 \log_{10} 250 + 0.52 \\
&= 1.19 \times 2.3979 + 0.52 \\
&= 2.854 + 0.52 \\
&= 3.374 \\
\therefore e &= 2366 \ \mu V
\end{aligned}$$

4.2.2. Peltier Effect

When a current is flowing through the junction, absorption or liberation of heat takes place, dependent on the direction of current flow. This is known as the *Peltier effect*.

4.2.3. Thomson Effect

A third effect, called the *Thomson effect*, may be stated as:
potential differences are set up in any given metal if it is not at the same temperature throughout.

Example.—The thermoelectric e.m.f. of a thermocouple is given by $e = 17.5t - 0.025t^2$, where t is the temperature in °C of the hot junction, the cold junction being at 0°C. Calculate the inversion temperature.

At the inversion temperature $e = 0$.
$$\begin{aligned}
\therefore 0 &= 17.5t - 0.025t^2 \\
0.025t &= 17.5 \\
t &= 17.5/0.025 \\
&= 700°C
\end{aligned}$$

The equation is a quadratic, therefore the curve is symmetrical about the neutral temperature and the inversion temperature is double the neutral temperature.

$$\therefore \text{neutral temperature} = 700/2 = 350°C$$

Alternatively:
$$\begin{aligned}
de/dt &= 17.5 - 2 \times 0.025t \\
&= 17.5 - 0.05t
\end{aligned}$$

For a turning-point $de/dt = 0$ = neutral temperature point

$$\begin{aligned}
\therefore 0 &= 17.5 - 0.05t \\
0.05t &= 17.5
\end{aligned}$$

$\therefore$ neutral temperature $\quad t = 17.5/0.05$
$\qquad\qquad\qquad\qquad\qquad = 350°C$
and inversion temperature $= 700°C$

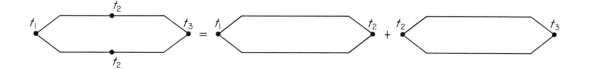

Fig. 48. The law of intermediate temperatures

4.2.4. Law of Intermediate Temperatures

The e.m.f. of a thermocouple with junctions at t_1 and t_3 is the algebraic sum of the e.m.f.s of two couples of the same metals with junctions at t_1 and t_2, and t_2 and t_3 respectively.

4.2.5. Law of Intermediate Metals

If one or more wires are introduced into the loop at the cold-junction end, the e.m.f. generated will be unaltered providing that the new junctions formed by introducing the extra wires are at the same temperature as the original junction.

A practical application is shown in Figure 51.

4.2.6. Thermoelectric Power

The rate of change of thermoelectric e.m.f. with temperature of one junction of a thermocouple while the other junction is kept at a fixed temperature is called the *thermoelectric power* of the couple and is a linear function of the temperature.

Lead is a metal that shows very little thermoelectric effect and is often used as a standard of comparison. Typical thermoelectric power lines for various metals, with respect to lead, are shown in Figure 49.

Figure 50 shows the e.m.f. generated by two metals with lead as their reference metal and junctions at temperatures t_1 and t_2. The thermoelectric power is represented by the area of the shaded portion of the graph. In (b) the neutral temperature is between t_1 and t_2; the total e.m.f. will then be the difference of the two shaded areas (because the e.m.f. reverses after the neutral temperature).

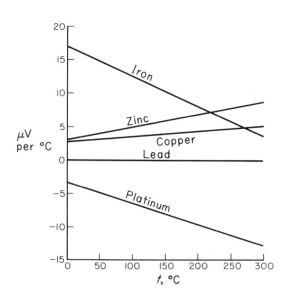

Fig. 49. Thermoelectric powers of metals in conjunction with lead

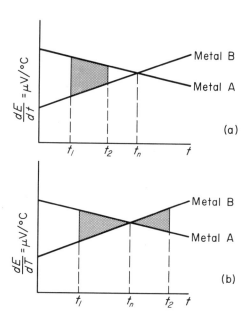

Fig. 50. Thermoelectric power

4.2.7. Thermocouple Leads

The simplest type of installation will consist of a thermocouple of two dissimilar wires with the hot junction immersed in the medium to be measured, and the cold junction connected to a very sensitive moving-coil galvanometer.

This arrangement is not always practical because of the cost of the extension wires. The thermocouple leads may terminate at a cold junction, near the hot junction, and be taken from there on by copper leads. In this arrangement the cold-junction temperature may vary considerably due to heat radiated from the heat source. It is not a method which is normally recommended.

To remove the cold junction from the hot zone, and to reduce the cost of extension leads, 'compensating leads' may be connected.

Compensating leads are made of material with similar characteristics to the thermocouple. The e.m.f.s generated where the compensating leads join the actual thermocouple should be equal and opposite; this will be so if both sides of the junction are kept at the same temperature and the correct leads have been chosen. These leads are almost always used with precious-metal couples because of the cost.

In the case of platinum and platinum/rhodium thermocouples, a copper and copper/nickel combination can be used, the copper being connected to the platinum/rhodium side. With chromel/alumel thermocouples, copper/constantan leads may be used, providing the thermocouple terminals are below 120°C.

When the length of compensating cable would be excessive, or inconvenient, the compensating cable is run to a convenient position for the cold junction where it may be maintained at constant temperature, the remainder of the wire being of copper.

Calibration is usually carried out with the cold junction at the ice point. If the cold junction is above the ice point, the e.m.f. is reduced by an amount corresponding to the temperature difference between 0°C and the actual cold-junction temperature.

4.2.8. Cold-junction Reference

The cold junction must be kept at a constant temperature, or some form of compensation for variation must be included. Several methods are available for maintaining constant temperature and include the following: thermostatically controlled oil baths, a thermos flask filled with melting ice, a pipe buried in the ground to a depth of about one and a half metres. A triple-point cell will maintain a temperature of 0·1°C for several hours.

4.2.9. Resistance of Thermocouple Circuit

Owing to small volt drops in the circuit, the voltage detected by a galvanometer will not be the true e.m.f. generated by the thermocouple. Resistance is made up of thermocouple, leads and indicator. If the resistance of any part of the circuit changes, for example, due to ambient temperature change, then there will be an error in the temperature indication. Because of this, in multipoint installations where switching takes place, gold- or silver-leaved contacts are often used.

If R_C, R_L and R_G are the resistances of the couple, leads and galvanometer respectively, then from Ohm's law, the current i will be

$$i = \frac{e}{R_C + R_L + R_G}$$

and the p.d. across the galvanometer v will be

$$v = iR_G$$
$$= \frac{e}{R_C + R_L + R_G} R_G$$

From this equation it will be seen that the greater the resistance of the indicator the nearer the reading will be to the true generated e.m.f. If the indicator resistance is high enough, then it may not be necessary to apply a correction for ambient variation.

Example.—A chromel/alumel thermocouple produces an e.m.f. of 41·3 mV at 1000°C, and together with its compensating leads has a resistance of 10 ohms. Calculate the value of series, or ballast, resistance required with the above thermocouple so that a galvanometer of 40-ohms resistance and full-scale deflection 0·2 mA may be used.

What voltage will be indicated on the galvanometer?

From Ohm's law $R = V/I = 41·3/0·2$
$= 206·5$ ohms

(this is the total resistance)

$$R = 10 + 40 + R_b$$
$$206·5 = 50 + R_b$$
$$\therefore R_b = 206·5 - 50$$
$$= 156·5 \text{ ohms}$$

Voltage across galvanometer $= IR_G$

$$= 0·2 \times 40$$
$$= 8 \text{ mV}$$

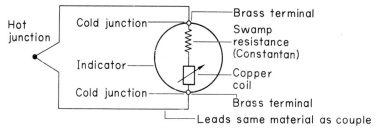

Fig. 51. Simple thermocouple installation

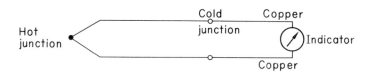

Fig. 52. Copper extension leads

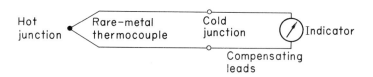

Fig. 53. Compensating leads

Fig. 54. Arrangement for long lengths

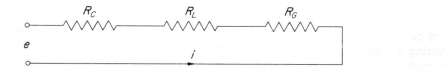

Fig. 55. Equivalent circuit of a thermocouple

4.2.10. Cold-junction Compensation

When it is not easy to keep the cold junction at a fixed temperature, compensation can be made. It may be achieved by using a bimetal strip as in the mercury-in-steel thermometer.

The bimetal strip will deflect linearly, but if the thermocouple characteristic is non-linear, full compensation may not be achieved.

4.2.11. The Millivolt Indicator

Essentially the indicator is a moving-coil millivoltmeter, but there are several points of difference between the instrument used for temperature measurement and that used for general electrical measurements. One of the major differences is that for temperature measurement the coil resistance is usually much higher. An accuracy of $\pm 1\%$ of full scale is achieved.

4.2.12. Errors and Corrections

Errors may occur due to (*a*) leakage currents and induced e.m.f.s from neighbouring electrical circuits; (*b*) breakdown of insulation of the refractory material of electric furnaces at high temperatures; (*c*) electrolytic effects due to dampness at any metal junctions, including switches; (*d*) thermoelectric effects due to the existence of temperature differences between junctions of dissimilar metals other than in the thermocouple itself.

Errors due to (*a*) and (*b*) may sometimes be eliminated by suitable earthing and shielding. (*c*) is mostly observed under damp conditions in which condensation may take place on any part of the circuit. (*d*) frequently results from air currents cooling one part of the circuit, or from unequal heating of the circuit due to the presence of local sources of heat radiation.

The instrument itself may be affected by stray magnetic fields unless it is sufficiently shielded.

4.2.13. The Potentiometric Circuit

For high accuracy a potentiometer is used to measure the actual e.m.f. of a thermocouple. When an unknown e.m.f. is applied in opposition to a sufficient supply e.m.f. of known value across a slidewire, at some point along the slidewire the two e.m.f.s will be equal. No current will flow through the galvanometer, the potentiometer being balanced at this point.

Putting a high resistance in series with the slidewire enables balance to be obtained over a reasonable length of the slidewire (Figure 59). The addition of a standard cell ensures that the correct potential drop is maintained across the slidewire. The standard cell is switched in at periodic intervals

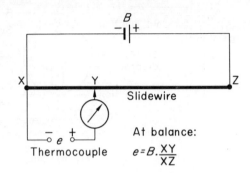

Fig. 58. Basic potentiometer

At balance:
$e = B \cdot \dfrac{XY}{XZ}$

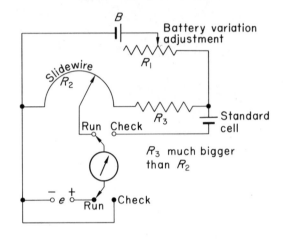

Fig. 59. Modified potentiometer

and R_1 adjusted until the galvanometer reads zero.

Compensation for variation of cold-junction temperature must be included if the cold-junction temperature is not held constant. The adjustment may be manual or automatic.

R_4 is wound with nickel wire, and the potential across it is arranged to be equal to that across R_6 at 0°C. As the resistance of R_4 varies with ambient temperature, so does the potential across it, thus the sum total of the thermocouple e.m.f. and e.m.f. across R_4 remains constant. By suitably choosing R_5, the current can be made to give a change in p.d. exactly equal to the thermocouple output for the same change in temperature.

4.2.14. Differential Temperature Measurement and Multipoint Installations

Temperature differences may be measured by connecting two similar thermocouples in series opposition. Multipoint installations may be connected in a similar manner to that for the resistance thermometer. The main difference is that the thermocouple voltages generated will be very much smaller

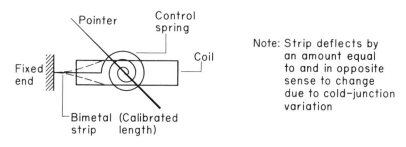

Fig. 56. Bimetal cold-junction compensator

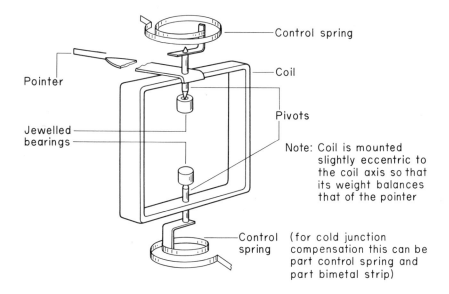

Fig. 57. Moving-coil system (Foster Instrument Co.)

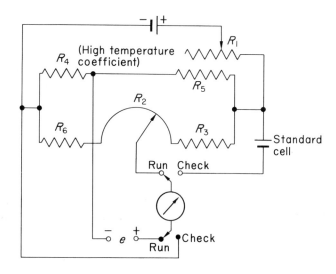

Fig. 60. Potentiometer with automatic cold-junction compensation

than those used in resistance thermometer installations, and switch contact resistance will be of greater importance. Good contact is essential throughout the whole circuit.

4.2.15. Choice of Thermocouple

In general thermocouples may be divided into two groups: rare metals and base metals. Base metals are used for temperatures up to about 1200°C, rare metals up to about 1500°C. Certain non-metal thermocouples may be used at temperatures beyond the range of the rare metal group. Base metals develop relatively high e.m.f.s though they oxidize at lower temperatures than rare metals. Rare metals are costly and their characteristics are less linear than those of the base metals.

4.2.17. Calibration

As mechanical strain may affect the characteristics, annealing should be carried out before a test is made. With a rare-metal thermocouple which has already been in service, re-annealing should be carried out by heating to a temperature of 200°C for one hour. For maintenance tests, it is usually sufficient to test at one standard temperature, since it is generally true that any alteration in the shape of the calibration graph affects the value of the e.m.f. at all temperatures. With a 'freeze test' (sec-

4.2.16. Thermocouple Materials

Table 3.   Industrial thermocouple metals

	Positive Element	Negative Element	Maximum Continuous Temperature, °C	Intermittent Temperature, °C	Notes
BASE	Copper	Constantan	400	500	
	Chromel	Constantan	700	1000	Generates a higher e.m.f. for a given temperature than any other type
	Iron	Constantan	850	1100	Iron element deteriorates rapidly at high temperatures due to scaling in oxidizing atmospheres
	Nickel/chromium (Chromel)	Nickel/aluminium (Alumel)	1100	1300	Susceptible to attack by carbon, sulphur and cyanide fumes. Better in oxidizing than in reducing atmosphere
RARE	13% platinum/rhodium alloy	Pure platinum	1400	1600	Must be protected against attack by metallic vapours and furnace gases; low e.m.f. generated
	Platinum/20% rhodium	Platinum/5% rhodium	1500	1700	Must be protected against chemical attack at high temperatures
	Tungsten	Tungsten/20% rhenium	2500	2800	For use in reducing or inert atmosphere only. Special insulation is necessary to limit chemical attack. High e.m.f.

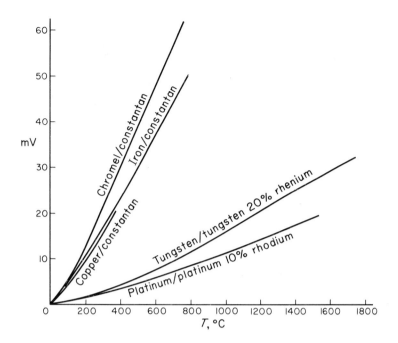

Fig. 61. Temperature/millivolts of industrial thermocouples

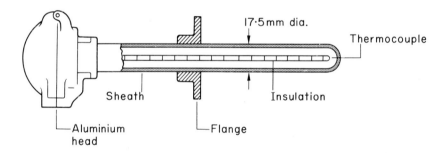

Fig. 62. Thermocouple element (pyrometer)

tion 6.2) of a standard substance, errors will occur if there is insufficient mass of the test substance, insufficient immersion or too close proximity to the bottom of the containing vessel.

4.2.18. Installation

Conductors should be small to reduce thermal conduction, radiation and response time; conversely, for mechanical strength, the cross-section should be large. In practice, with base metals, for spot readings up to 900°C the wires should be not less than 1·6 mm diameter. For higher temperatures and continuous readings 3·25 mm is common. With rare metals, mainly because of high cost, 0·5 mm is used successfully.

Insulation between the conductors except at their hot junction is essential and, except in special cases, they should be insulated from earth. The insulation material will be determined by the temperature range, and it is usually essential, unless working below 500°C, to protect the thermocouple with a sheath.

4.2.19. Automatic Self-balancing Systems

Because of their inherent accuracy, it is often desirable to operate bridges and potentiometers in the balanced condition. When balanced there is no current flow through the detector, and hence no errors due to volt drops. Variation of supply voltage will affect the system sensitivity, but there is generally a visual indicator to give the level of the supply voltage. The principle of operation is similar for both bridge and potentiometer. A change of temperature will produce an 'error' which will cause an unbalance current to flow through the circuit diagonal, the error is amplified and causes a balancing motor to alter the position of the sliding contact of the bridge or potentiometer in such a manner as to reduce the error to zero and rebalance the circuit.

Most modern designs use semiconductor circuits, the only moving parts being the balancing motor, the shaft of which is usually coupled to the slider and to the pen or pointer.

4.2.20. Commercial Types of Measuring Instruments. The Tenzor

The Tenzor system, manufactured by Negretti and Zambra, may be used for either thermocouples or resistance thermometers.

Any change in the value of R_t, due to change of temperature, causes an out-of-balance signal to appear at the bridge output. This is fed to the d.c. amplifier, and appears as a change in output heating current through the tube.

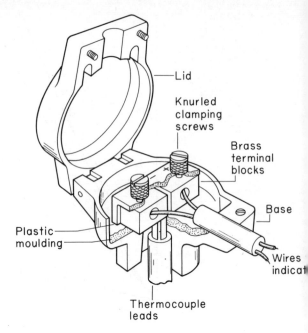

Fig. 63. Typical thermocouple (pyrometer) head (Cambridge)

Considering a rise of output, the increased current has two effects: (a) it causes further expansion of the tube and thus, through the linkage, a new deflection of the pointer or the pen; (b) it increases the resistance R_b of the tube winding, thus tending to rebalance the bridge and reduce the output signal to the amplifier. The heating current in the tube then reduces to a level sufficient to maintain the tube at this new temperature, and an equilibrium will again be established. The accuracy claimed is $\pm 1\%$.

4.2.21. Foster Potentiometric Recorder

The Foster instrument is more conventional than the previous instrument. The circuit is similar to Figure 60 but the galvanometer is replaced by a chopper amplifier. (The chopper converts the d.c. signal to a 50-Hz modulated carrier wave.)

The direction of rotation of the balancing motor depends on the polarity of the chopper signal which varies the phase of the amplifier output by 180° depending on whether the out-of-balance voltage goes positive or negative with respect to its value before the change commenced.

4.2.22. The Fielden 'Bikini'

The Fielden 'Bikini' recorder is an example of yet another method of null balance recording.

When voltages V_1 and V_2 are equal, the bridge is in balance; current will flow through transistors Tr_1 and Tr_2 on alternate half-cycles, until C is

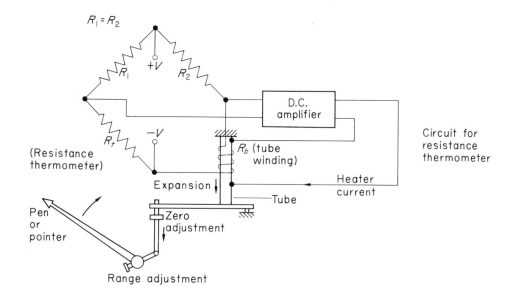

Fig. 64. Negretti and Zambra 'Tenzor' (resistance thermometer)

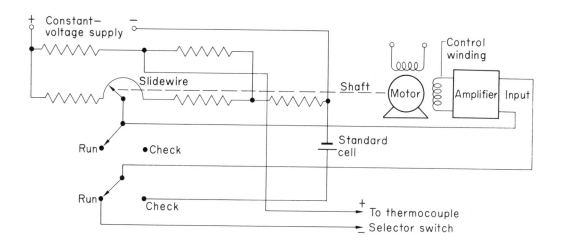

Fig. 65. Foster 3670 recorder (simplified diagram)

Table 4. Summary of electrical thermometers

Measuring element	Application	Range, °C	Remarks
RESISTANCE			
Metal	Precision measurement	−240 to 600	Generally used for low temperatures giving good accuracy
Photo-etched	Surface temperatures	−200 to 300	May need the use of a strain-gauge detector. Requires careful fitting
Thermistor	Laboratory use	−100 to 300	Cheap elements but may produce non-linear scale. Stability may be unreliable
THERMOCOUPLE			
Base metal	General industrial use	−200 to 1150	Normally used except for higher temperatures
Rare metal	Precision measurement	0 to 1450	Expensive. Compensating cables usually required

fully charged to the level equal to the difference between the amplifier input bias and V_1 or V_2 when no more current will flow.

If the thermocouple voltage rises, it appears as an unbalance across the bridge. V_1 will now differ from V_2 and C will be alternately charged through one transistor and discharged through the other; thus a.c. will flow through the amplifier's input circuit, and the phase and amplitude of this current will depend upon the direction and amplitude of the d.c. appearing as unbalance across the bridge.

Thus, if the unbalance is in one direction, the amplifier will receive an increase of input, and if the unbalance is in the other direction, it will receive a reduction of input. Thus the motor will rotate in a direction appropriate to the restoration of the balance of the bridge.

The rebalance accuracy is as good as the slide-wire resolution, of the order of 0·1%.

Examination Questions

1. Describe, with the aid of a sketch, the construction of a resistance type of temperature-sensing device. Name *two* important properties of the material used for the resistance element.

Draw a diagram showing a typical resistance-thermometer installation and explain how errors due to change in the temperature of the connecting leads are reduced.

C.G.L.I. (310/3 1967)

2. A muffle furnace used in the heat treatment of metals is to operate up to a temperature of 1400°C. The furnace is fitted with a thermocouple which is connected to a millivoltmeter fitted with cold-junction compensation and located in the same room as the furnace.

(a) Suggest suitable materials for the thermocouple and compensating leads.
Explain why the latter are necessary.

(b) Describe fully, with sketches, the main constructional details of the millivoltmeter indicator.

C.G.L.I. (310/1/03 1968)

3. With the aid of a circuit and other diagrams describe the operating principle of an electronic multipoint self-balancing potentiometric recorder. Give details of any routine checks and maintenance that should be carried out on such a recorder.

C.G.L.I. (079/1/03 1968)

4. The figure shows a Wheatstone bridge arrangement for temperature measurement by means of a resistance thermometer.

Given that $X = Y = Z = 100$ ohms, $S = 102$ ohms, $r = 1$ ohm, $R_T = 135$ ohms at 20°C and that the resistance temperature coefficient is 0·004 ohm/ohm at 20°C, determine the temperature range which can be measured by this arrangement.

(± 140°C)   U.L.C.I. 1968

5. (a) The figure illustrates the basic measuring circuit of a null balance potentiometer. State the purpose of the nickel spool N and the resistors L and S. Describe, in terms of circuit voltage, how the nickel spool carries out the function for which it has been included.

(b) For the circuit shown calculate the value of the thermocouple e.m.f. to be injected at X X so that the detector will be just balanced at zero when the value of resistor S between B and C is 108 ohms.

(10 mV)   U.L.C.I. 1969

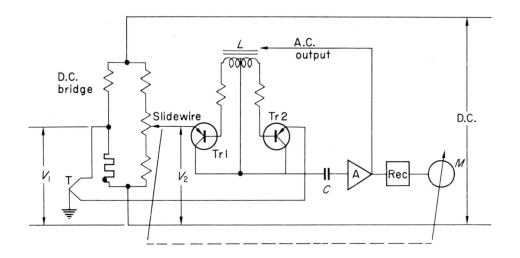

Fig. 66. Fielden 'Bikini' recorder

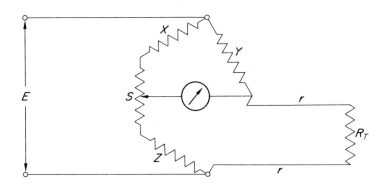

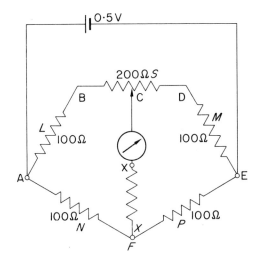

PART 5 — Radiation pyrometers

Radiation pyrometers are mainly used for temperatures above 1400°C, though they are capable of responding to temperatures as low as 500°C. They can make use either of visible light or of the whole spectrum of radiation coming from the body. The radiation coming from the bodies at relatively low temperatures may be entirely within the invisible infra-red bands.

5.0.1. Prévost's Theory

If two bodies are at different temperatures there will be a transfer of energy from the hotter to the cooler body. If both bodies are at the same temperature and enclosed in a space impervious to heat, each body will continually radiate heat into and absorb heat from the surrounding medium. The two processes produce equilibrium, so that the temperature of each body will remain the same.

5.0.2. Black-body Conditions

When a body at a given temperature radiates energy from its surface, the condition and colour of the surface are important. A *black body* is one which radiates the maximum possible energy at this temperature.

All ordinary bodies fall short of the ideal, so that it may be necessary to apply a correction factor defined thus:

Emittance factor
$$= \frac{\text{total radiation from a non-black body}}{\text{total radiation from a black body}}$$

assuming the two bodies are otherwise identical.

A black body is also a perfect *absorber* of energy. A non-black body absorbs part of the energy that falls on it and reflects the rest. The emittance factor is the same whether the body is radiating or absorbing.

It is fortunate that black-body conditions apply when viewing a body in a furnace through a small hole. Radiation reflects from the furnace walls on to the body and finally out through the viewing hole. Provided the temperature is uniform, and there are no confusing effects from the flames, the radiation pyrometer is a reliable instrument for measurements of furnace temperature.

5.0.3. Stefan-Boltzmann Law

The total energy emitted from a black body into free space is proportional to the fourth power of the absolute temperature.

$$E = a\sigma T^4$$

where a is the area of the body, T is its absolute temperature and σ is an absolute constant $5 \cdot 67 \times 10^{-8}$ $Wm^{-2} K^{-4}$. Where a body is radiating to its surroundings at temperature T it receives a back flow of energy from them so that the expression becomes $E = a\sigma(T_1^4 - T_2^4)$. In practice T_2^4 can often be neglected.

Pyrometers which respond to all wavelengths, thus obeying the Stefan-Boltzmann law are called *total radiation pyrometers*. Instruments that utilize only a narrow band of wavelength in the visible region are called *optical pyrometers*.

5.1. Total-radiation pyrometers

Energy is radiated from the hot body and is focused on to a sensing element which rises in temperature until its heat loss to its surroundings balances the radiant heat it receives. The problem is then one of measuring the comparatively low temperature of the sensing element (about 400°C when the radiating source is at 1600K). The element may be a thermocouple or a phototransistor which is sensitive to radiation.

5.1.1. Lens Material

Glass is not always used for lenses because it may cut off some of the wavelengths of the radiation. Pyrex, fused silica, calcium fluoride and arsenic trisulphide are commonly used to extend the range of wavelength transmitted.

5.1.2. Fixed-focus Tubes

A mirror or lens is used to focus the radiation. In the mirror type, radiation falling within the cone of vision of the receiving tube impinges upon a concave mirror and is focused on the sensor.

Theoretically the action is independent of distance, provided the hot body fills the cone of vision of the tube. In practice, however, the reading does vary a little with distance, due to the effects of water vapour, carbon dioxide, convection currents from source to instrument and the effects of stray reflections within the receiver. A common relationship

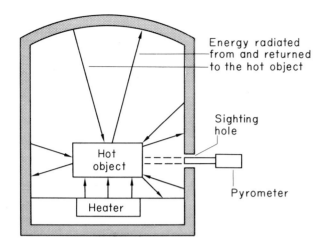

Fig. 67. Black-body conditions in a furnace

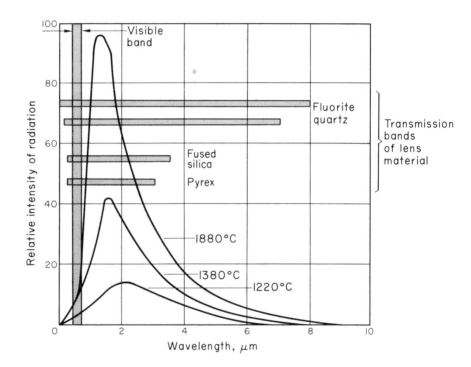

Fig. 68.
Spectral distribution of energy

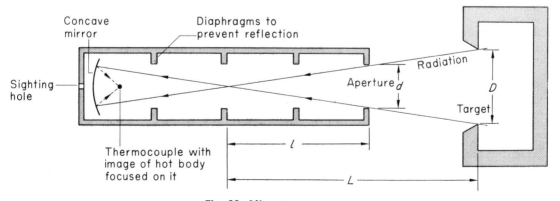

Fig. 69. Mirror-type pyrometer

between source diameter (D) and distance (L) is to make $L = 20D$.

The lens type is usually shorter in length but bigger in diameter. It is intended for fixed installations.

The receiving tube must be used at a prescribed distance. The maximum distance factor with the Foster pyrometer is 20:1. This means that the maximum distance the lens may be away from the hot body, or the inner edge of the orifice, is twenty times the diameter or smallest dimension of the orifice of the hot body. About ten seconds is required to obtain a steady reading, although 90% of a temperature change can be detected in 1 to 1½ seconds.

If it is not possible to have a permanent opening in the furnace wall, a closed-end tube of refractory material is inserted and attached to the open end of the tube.

In some installations it may be necessary to surround the pyrometer tube with cooling water.

If cold-junction compensation is required, a nickel resistance coil, or a thermistor, is fitted at the cold-junction end in such a manner as to compensate for any change in ambient temperature.

A correction factor may have to be applied if black-body conditions do not exist due to the presence of radiation-absorbing gases, smoke or luminous flame coming between the hot body and the pyrometer. Lenses and mirrors must be kept clean or incorrect results will be obtained.

5.1.3. Varying-focus Tubes

The Féry type of pyrometer uses a telescopic tube to enable the source to be focused on the sensing element.

A circular image appears when in focus, but the image is split when out of focus. Because it has to be focused it is unsuitable for recording or control.

5.1.4. Extension of Range

The range of a pyrometer may be extended by reducing the admitted radiation by a known amount. This may be achieved by means of a filter or by means of a rotating disc.

The disc has a sector cut out of angle θ radians, so that the radiation entering is reduced in the ratio $\theta/2\pi$. The pyrometer then indicates a temperature T_2 which is less than the true temperature T_1 of the source.

$$\therefore \frac{\theta}{2\pi} = \frac{T_1^4}{T_2^4}$$

whence $T_2 = T_1 (2\pi/\theta)^{\frac{1}{4}}$

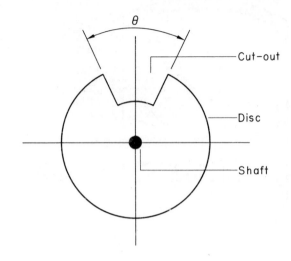

Fig. 72. Rotating disc for increasing range

5.2. Optical Pyrometers

Optical pyrometers make use of a single wavelength, or very narrow band of wavelengths, in the optical part of the spectrum for measuring the temperature of a hot body. The wavelength is usually chosen to be about $0.65\mu m$, in the red portion of the spectrum. Light entering the instrument passes through a glass filter so that it is almost monochromatic at the chosen wavelength.

5.2.1. Disappearing-filament Pyrometer (Variable-intensity Type)

The commonest form of optical pyrometer is the disappearing-filament type, in which the filament of a lamp blends into the illumination from the hot object when the illumination of both is the same.

When sighted, the lamp filament appears superimposed against the hot mass. By adjusting the variable resistance, the brilliance of the lamp is matched to that of the object, whereupon the filament becomes invisible. The transition points are quite definite, and different operators should obtain very similar results.

The lamp is never operated above 1450°C so that its calibration remains constant over long periods. Typical ranges are 800 to 1250°C with an accuracy of ±5°C, or 1100 to 1900°C with an accuracy of ±10°C. Because they have to be manually adjusted to obtain a reading, these pyrometers are unsuitable for recording or control.

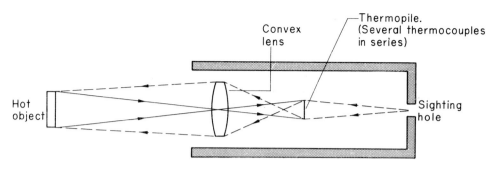

Fig. 70. Lens-type pyrometer

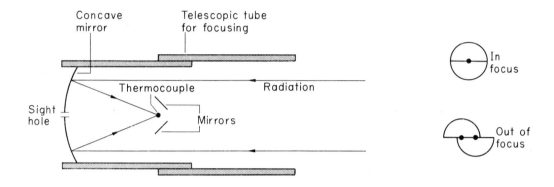

Fig. 71. Féry-type pyrometer

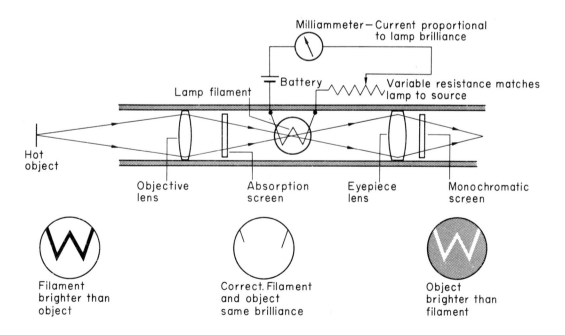

Fig. 73. Disappearing-filament pyrometer

5.2.2. Wedge-type Pyrometer (Constant Intensity)

The principle of the wedge-type pyrometer is that an optical wedge, or other type of attenuator, is situated between the source of radiation and the lamp, whose brilliance remains fixed. The wedge is adjusted until the light from the source is just extinguished, and the position of the wedge is then a measure of the temperature.

5.3. Photoelectric Pyrometers

The photoelectric pyrometer is similar in principle to the total-radiation pyrometer, except that the sensing element is a photoelectric cell which measures radiation falling on it within a comparatively narrow band of wavelength. They have a high response speed making them useful for measuring moving objects.

5.3.1. Fibre-optic Methods

In this system use is made of a special infra-red transmitting glass combination in the form of a fibre, used in bundles of approximately 4000, and shown in section in Figure 74.

Light entering at one end is totally internally reflected to the other end where the emergent infrared rays are sensed by, say, a phototransistor. Advantages are high speed of response and the ability to magnify or attenuate a signal as required. One application is the individual monitoring of the temperature of gas-turbine blades.

Examination Questions

1. Explain what is meant by the terms:
   (*a*) black body,
   (*b*) emissivity.

Describe, with the aid of a diagram, the construction and principle of operation of an industrial total-radiation pyrometer.

State the factors that could affect the accuracy of a radiation pyrometer and explain how these factors could influence the selection and installation of such a device.
   C.G.L.I. (310/4 1967)

2. Describe the construction and principle of operation of an industrial 'total' radiation pyrometer. State the sources of error with such a device and describe the methods adopted to reduce them to a minimum.

Describe briefly how the accuracy of calibration of a radiation pyrometer could be checked.
   C.G.L.I. (79/3 1967)

3. With the aid of diagrams describe the construction and principle of operation of the following type of disappearing-filament optical pyrometer:
   (*a*) constant-intensity comparison type,
   (*b*) variable-intensity comparison type.

Describe how the calibration of an optical pyrometer would be checked.
   C.G.L.I. (310/2/04 1968)

4. Describe the construction and principle of operation of a 'total' radiation pyrometer suitable for use in an industrial environment such as a steelworks.

State possible sources of error associated with such a device and describe how they are reduced or eliminated by the manufacturer in the design of the instrument and/or by the user in the correct application and maintenance of the pyrometer.
   C.G.L.I. (310-2-04 1969)

5. (*a*) Write an account of the optical pyrometer as a temperature-measuring device.

(*b*) A black-body furnace is heated until its temperature is 1227°C. If the diameter of the furnace is 60 mm, calculate the value of the radiant energy emerging from the furnace inlet, given that Stefan's constant $= 5 \cdot 735 \times 10^{-12}$ $J/cm^2/T^4/s$.

What would be the effect of decreasing the absolute temperature to half its original value?
   (820 J/s. 50·9 J/s.)   (B.I.T. 1969)

Table 5. Summary of radiation pyrometers

	Type	Range	Remarks
TOTAL RADIATION	Fixed focus	500°C upwards	Suitable for recording and control
	Féry	500°C upwards.	Unsuitable for recording or control
OPTICAL	Variable intensity	800°C upwards	Generally used for laboratory work and for spot checks
	Fixed intensity	800°C upwards	Neither of these are suitable for recording or control
	Photocell	800°C upwards	Suitable for measuring the temperature of moving objects

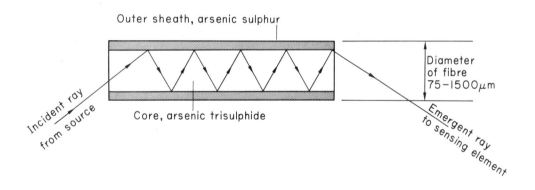

Fig. 74. Section through a fibre showing passage of ray

PART 6 Calibration and testing methods

Some of the more usual methods of testing and calibrating industrial thermometers are given below, these being based on B.S. 1041:1943.

6.1. Fixed-point methods

6.1.1. Ice and Steam Points

The ice and steam points are frequently used for calibrating thermocouples, resistance thermometers, mercury-in-glass and other thermometers. The ice should be finely powdered and moistened with water, and may be contained in a vacuum flask. The depth of immersion of the instrument should be sufficient to ensure that the testing body of the instrument has reached the same temperature as the melting ice. For high accuracy, the ice should be made from distilled water, and should be contained in a porcelain or other vessel provided with a hole at the bottom so that any excess water may be drained off. The effect of change of atmospheric pressure on the melting-point of ice is extremely small, and may normally be neglected.

For the steam point the instrument should be tested in a hypsometer. The steam temperature depends on atmospheric pressure and the latter should be measured with a barometer during test. The steam temperature can then be obtained by calculation or from standard tables.

6.1.2. Sulphur Point

The sulphur point, i.e. the boiling-point of sulphur (444·6°C) under standard conditions, is used for the calibration of thermocouples and resistance thermometers. The test is carried out in a special apparatus which is usually designed in accordance with a specification recommended in the International Temperature Scale for testing resistance thermometers. Briefly, the sulphur is brought to the boil and the heat supply adjusted until the vapour level is clearly visible. The thermometer under test is then immersed in the vapour only.

Recent work favours the use of the freezing-point of zinc in place of the boiling-point of sulphur.

6.2. Freezing-point methods

The freezing-point method is most frequently used for the calibration of thermocouples. The chief apparatus required is a suitable furnace and a number of crucibles containing the freezing-point substances, similar to that shown in Figure 76. The depth of immersion of the thermocouple should be chosen so that there is no change in the e.m.f. of the thermocouple at the freezing-point if the thermocouple is raised or lowered by 1 cm from its normal position.

The usual procedure is as follows: The crucible of metal is heated to a temperature a little above the freezing-point, and then the thermocouple, protected by a suitable sheath, is inserted into the molten metal. The furnace is then allowed to cool, and the e.m.f. of the thermocouple is measured with a potentiometer. When the freezing-point is reached, the e.m.f. of the thermocouple should remain constant for a short time since it is a latent-heat point.

6.3. Melting-point of wire method

This method is most frequently used for the calibration of rare-metal thermocouples at the melting-point of gold, and it necessitates only a small amount of the melting-point material. The thermocouple is broken at the hot junction, and a short length of pure gold wire, about 0·2 mm diameter, is welded between the two ends of the thermocouple wires. The thermocouple is then heated in a furnace so that the hot-junction end is situated at a uniformly heated portion of the furnace, and so that there is a minimum amount of strain on the gold wire. The rate of heating is controlled so that the temperature rises very slowly, about 1–2° per minute as the melting-point is approached; and then the e.m.f. of the thermocouple is observed at frequent intervals. The e.m.f. immediately prior to the junction breaking is taken as corresponding to the melting-point. Care should be taken when carrying out this type of calibration as the galvanometer will receive maximum unbalance current when the wire melts.

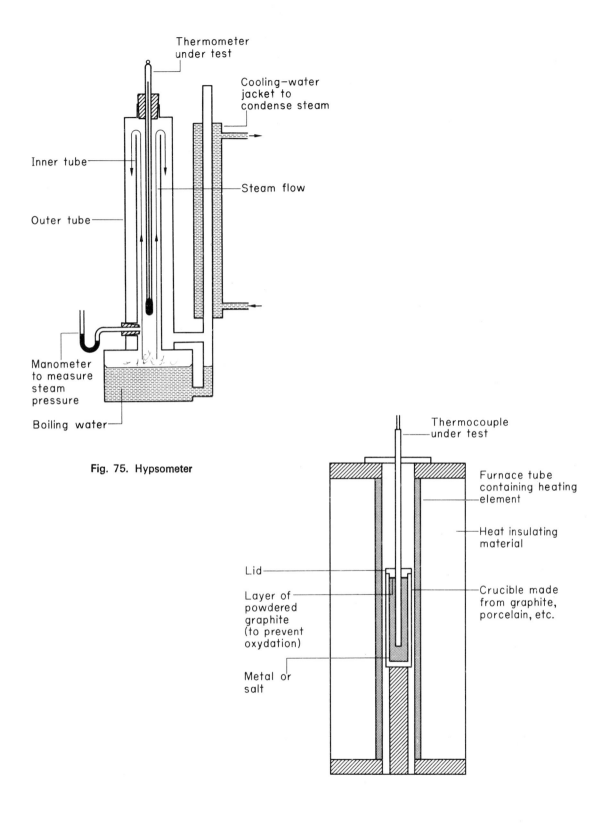

Fig. 75. Hypsometer

Fig. 76. Furnace for freezing-point method

6.4 Comparison methods

Comparison methods are frequently used for the calibration of most types of thermometers when a standard instrument is available. The liquids generally used are water 0–100°, oil 100–250°, salts or metal baths, or metal blocks 250–550°C.

6.5. Calibration above 1063°C

6.5.1. Secondary Standards of Calibration

The National Physical Laboratory undertakes the calibration of tungsten-strip lamps up to 2500°C and also calibrates optical pyrometers at the apparent temperature of the carbon arc, 3500°C. Similarly, furnaces operating under substantially black-body conditions, whose temperature can be measured by means of a calibrated thermocouple or optical pyrometer, may be used for the calibration of total-radiation pyrometers, providing that care is taken that the radiating area covers the whole field of the instrument.

6.5.2. Calibration of Radiation Pyrometers by the Melting-point of Refractory Cones

As mentioned earlier, certain refractory materials have a definite melting-point which is accurately repeatable when they are obtained in the pure state.

A small cone of the material is constructed and placed in an electric furnace consisting of a refractory tube, the greater part of which can be heated uniformly to a temperature desired. The furnace tube is open at both ends, so that, by looking straight along it, it is possible to see the cone clearly. On the other hand, by looking slightly diagonally, a good approximation to black-body conditions can be obtained. The furnace is brought approximately to melting-point, and then heated slowly 2–5°C per minute, until the cone is seen to collapse, and the corresponding reading of the pyrometer is taken at this point; this gives the calibration point.

City and Guilds Examination Questions

1. Describe the fixed-point and comparison methods of calibrating industrial temperature-measuring instruments over the range from the ice to the gold point.

Your answer should include details of the measuring equipment necessary and should contain well-drawn diagrams of such items as the furnaces and liquid baths that would be employed.
(310–2–04 1969)

2. Give an account of the fixed-point and comparison methods of calibrating thermometric elements in the range 0°C to 1063°C. Describe the furnaces, baths, and other equipment used, giving details of any special precautions necessary to ensure accurate, safe and effective operation.
(310/2/04 1968)

3. Make a list of the equipment necessary for setting up the temperature-calibration section of an instrument department.

Explain very briefly the principle of operation of each item and the reasons for its inclusion.
(310/4 1967)

4. Describe briefly what is meant by the International Temperature Scale.

Describe the fixed-point and comparison methods of calibrating and testing industrial temperature-measuring instruments over the range 0°C to 1063°C. Your answer should include brief details of the measuring equipment and the furnaces and liquid baths used.
(79/3 1967)

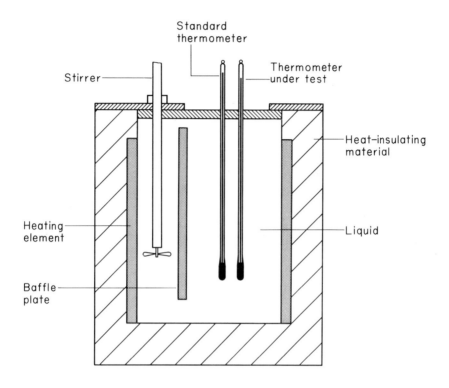

Fig. 77. Liquid bath for comparison tests

Index

absolute temperature 1
ageing 6, 8
ambient-temperature compensation 12, 24
automatic self-balancing 38

'Bikini' 38
bimetal compensator 12, 30
   thermometer 3
   thermostat 4
black body 42
bulbs 10, 12, 14

calibration 8, 38
   methods 48
Callender-Van Dusen equation 23
capillaries 11
City and Guilds examples 22, 40, 46, 50
class A certificate 7
cold junction 29
   compensation 34
   reference 32
comparison methods 50
compensating leads 27
crayons 2
critical temperature 18
cross ambient effects 18

differential temperature 29, 36
dip effect 13
disappearing filament 44

errors 34

fibre optics 46
filling liquids 8, 13
fixed-focus tubes 42
fixed-point methods 48
Foster potentiometric recorder 38
four-wire circuits 27
freezing-point methods 48
fundamental interval 23

gas-filled thermometers 16
gold point 1

head errors 18
hysteresis 13

ice point 1
installation 28, 38
intermediate metals, law of 31
   temperatures, law of 31
international temperature scale 1
inversion temperature 30

Kelvin scale 1

lead resistance 27

lens material 42
linearization 26
liquid-in-glass thermometer 6
liquid-in-metal thermometer 10

maximum thermometer 8
measuring circuits 26
melting-point of wire methods 48
millivolt indicator 34
minimum thermometer 8
multipoint installations 28

neutral temperature 30

optical pyrometers 44
oxygen point 1

paints 2
parasitic phenomena 25
Peltier effect 30
photo-etched elements 24
pipe runs 20
pockets 14
potentiometer circuits 34

range 14
refilling 13, 16, 18
refractory cones 50
regulators 14
remote transmission 20
resistance thermometers 23

saturated vapour pressure 18
secondary standards 50
Seger cones 2
self heating 24
sheaths 14
steam point 48
Stefan-Boltzmann law 42
sulphur point 1, 48

thermistor 25
thermocouple 29
   choice of 36
   materials 36
thermoelectric inversion 30
   power 31
thermostats 5
Thomson effect 30
three-wire circuits 27
time constant 24
total radiation pyrometer 42

varying-focus tube 44

wedge type 46

zero depression 7